MY FRIENDS
CALL ME C.C.

MY FRIENDS CALL ME C.C.

The Story of
Courtney Chauncey Julian

By

WILLIAM GARDINER HUTSON

Sunstone Press
Santa Fe, New Mexico

The photograph of C.C. Julian is from the July 14, 1931 issue of the "Oklahoma City Times," Copyright 1931, The Oklahoma Publishing Company.

The author wishes to thank the Hathaway Ranch Museum, Santa Fe Springs, California 90670 for historical documents concerning C.C. Julian and also Cultural Horizons, Pomona, California 91767 for manuscript production.

Printed in the United States of America

Library of Congress Cataloging in publication data:

Hutson, William Gardiner, 1928 -
        My friends call me C.C. / by William Gardiner Hutson.
                p.   cm.
        ISBN 0-86534-143-5
        1. Julian, Courtney Chauncey. 2. Industrialists -- United States -
-Biography. 3. Petroleum industry and trade -- United States -
-History. 4. Petroleum industry and trade -- Canada -- History.
I. Title.
HD9570.J85H88   1990
338.2'728'092 -- dc20
        [B]   90-35789
                                                                    CIP

Published in 1990 by SUNSTONE PRESS
                Post Office Box 2321
                Santa Fe, New Mexico 87504-2321 / USA

My Friends
Call Me C.C.

Courtney Chauncey Julian

< 7 >

CHAPTER 1

The Santa Fe conductor on the Chicago to Los Angeles train for his midnight run from Seligman, Arizona to Barstow, California, alerted passengers in the Pullman car: "Barstow, next stop Barstow! We'll be there one hour. Passengers departing pick up your luggage inside the station. We arrive in ten minutes at 7:35 a.m. Pacific time. Breakfast will be served at the Fred Harvey House!"

Chauncey Julian slid his gold poker watch from his vest pocket, checked the time, and advanced the hands one hour to conform to Pacific time. He had been sleeping soundly when the West-bound train crossed the Colorado River on the steel cantilever bridge connecting Arizona and California near Needles.

Barstow came into view through the southerly car windows—gray boulders, Jerry built shacks, and dozens of idle box cars. Moments later, the engine hissed and shuddered to a halt west of a huge willow tree which partially shaded the Barstow Santa Fe Station from the summer sun.

Chauncey Julian, tall, muscular, appearing self confident, stepped down from the Pullman car to the gravel area leading to the station house. A fresh set of train men, conductor, engineer, and fireman stood in a group preparing for the final run to Los Angeles.

The desert air, usually dry, was heavy with a haze which the morning sun, like the proverbial Indian giver, wafted up into the atmosphere from the residue of shallow puddles of rain water, the gift of a desert storm which had blown in and away just before sunrise.

Striding casually toward the railway station, Chauncey Julian's attention was diverted toward a crowd of people seated at dining tables within an ornate two-story building with granite pillars like a county courthouse. Above the portals of the antechamber to the dining room was a sign in Spanish, "Casa de Desierto."

He entered the station-house to the west of the Casa de Desierto, which was the Fred Harvey dining area, giving a boost with his boot to the polished brass kick plate at the bottom of the heavy wooden door.

To his left, on either side of the baggage room Dutch door, were two neat stacks of morning newspapers brought from Los Angeles on the early eastbound train. Tossing a nickel into a coffee can with a slit in the lid, Chauncey Julian picked up a copy, seating himself on the hard wooden bench which ran parallel across the waiting room. Scanning page one, he then flipped pages to the business section, searching for a news item about his favorite subject, the oil industry, which was "booming," an unfavorable word in and around Los Angeles. He sat upright! His attention was focused on one news item:

SIGNAL HILL BEACON OF SUCCESS
FOR OIL INVESTORS,
Feb. 24, 1922, Long Beach, CA. Frederick Greenleaf,
President of recently incorporated Arrowlube Oil

< 8 >

Company, announced the public sale of 400,000 shares of common stock. Greenleaf expressed optimism that the public's acceptance of Arrowlube gasoline products would triple output requirements at the company plant near Signal Hill. Greenleaf added; "Potential investors in our stock are welcome at our   plant near Signal Hill from 9:00 a.m. to 5:30 p.m., Monday through Friday." He continued; "Our stock will be traded on the Pacific Coast Stock Exchange or may be purchased directly from our company at a clearly marked tent near Signal Hill."

During a meeting with reporters at the Spud-In Cafe on Signal Hill, Mr. Greenleaf disclosed that his company had entered an agreement with Western Star, Hoot Gibson, to promote Arrowlube gasoline at his rodeos and also on radio. Mr. Greenleaf believes his company can quickly add to its share of the gasoline market in California.

Chauncey Julian folded the paper, tossed it on the bench, leaned back, clasped his hands behind his head, stared at the ceiling, and sat motionless as if mesmerized.

"You seem to be in a dream world, Sir." Julian lurched forward, dropping his hands to the wooden bench. He had been oblivious of a fellow passenger; "Good morning, Sir, Feeney is the name, William Warren Feeney." Julian shook hands without rising, "Sit down, do they call you Bill?"

"Sure do."

"You caught me, Bill. I'm a born dreamer." Chauncey Julian retrieved the newspaper at his side, opened it to the business section and pointed to the article about the oil company active on Signal Hill. "That news item brought me a great deal of pain. Last year, things did not go well for me. The first part of the year I sold oil stock for a fellow I met when I was working in the oil fields in Texas. I sold the stock easily, mainly around Los Angeles. Also, sold to some folks I know in Bakersfield, worked there as a rigger in '08, but all four of the Texan's wells came in dry. To top that, I scraped up some money and drilled on a small lease I had in Huntington Beach and it came in dry. This year, things will be different. I'm planning my own oil company— going public as they say. By the way, I'm Chauncey Julian, Courtney Chauncey Julian to be exact, my friends call me C.C. What should I call you?"

"Depends."

"Are you into oil?"

"Do I look like an oil man?"

"Not the ones I've been working with. You look like an Easterner— New York?"

"You're right, C.C. Just graduated from Columbia School of Journalism. I was hired by the Hollywood Citizen News. They tell me the circulation is mainly in Hollywood, and some place they call the Valley. They need a reporter to cover City Hall in downtown Los Angeles. I've rented a room at a small hotel on Wilcox near the paper's offices."

< 9 >

"Well, good luck to you, Bill. You're moving into a town where there are two types of folks. In one group there are the hustlers, and in the other group they call themselves respectable. One group wants to make money fast and the other group has money and keeps it to themselves. This go-around in Los Angeles, I'm going to do business my way and I'm going to make a lot of money fast." "Well, C.C., I'm no hustler. My father is part of the established group—a newspaper publisher. My ambition is really not to make a lot of money or to be wielding power. Truthfully, I want to settle somewhere in the Los Angeles area, find the right woman, raise a family, and support the wife and family as a line reporter. Maybe someday, I'll return to the east coast, I'm not really sure, there are a lot of options."

Before Chauncey Julian could respond, the door of the waiting room was flung open, so as to startle the station master and ticket agent who were busy with paperwork behind the half-open Dutch door. A rotund cowboy strode up to the ticket agent. He wore a beaver Stetson tilted forward and cocked to the right. His navy blue suit was newly pressed. His white shirt adorned with a blazing red tie, partially covered a paunch upheld by a silver buckled cowboy belt. Smoking a corncob pipe, he approached the ticket agent and took a large wad of greenbacks from his coat pocket.

"One round-trip to Los Angeles, Frank."

"Yes sir, Mr. Scott!"

The ticket agent quickly stamped the ticket and handed it to the cowboy with deference. He nodded his gratitude to the ticket agent and turned facing Chauncey Julian and Bill Feeney, who remained seated watching the ticket transaction.

"How are you boys doing...I'm Death Valley Scotty."

Julian rose and shook the cowboy's outstretched hand. Bill Feeney did likewise stammering, "I heard of you, Mr. Scott...I mean when I was in grammar school, my parents talked about your extra fast trip on the Santa Fe Railroad from Los Angeles to Chicago."

Death Valley Scotty grinned approval at the recognition. "What's your handles, boys?"

"I'm Bill Feeney, and this gentleman is Chauncey Julian. I'm a newspaper reporter and Mr. Julian is an oil man."

"Pleased to meet you, Beeney, and you too, Chancey."

"That's Chauncey, Mr. Scott."

"My friends call me Scotty."

Death Valley Scotty turned to Bill Feeney, winked, then, addressed Julian: "After black gold, Chancey? I like solid gold. There's lots of it out there." Death Valley Scotty made a sweeping gesture with his right hand toward the northeast. "My castle is up there—Death Valley Ranch. You boys should head up there one of these days. Beeney, I can tell a story or two for your readers. And as for you Chancey, if you don't find any black gold, I can set you off looking for some real gold...lots of it out there, I've been told."

Julian placed his thumbs in his vest pockets and looked seriously at Scotty. "Mr. Scott, I have acquired a leasehold of extremely valuable oil property in Santa Fe Springs. I may, also, at a later date venture into mining, not gold—lead. I'm told there may soon be a worldwide shortage of lead.

< 10 >

While in Arizona, I met some gentlemen in the lead mining business and I am negotiating to participate in a claim."

Death Valley Scotty seemed not to take Julian seriously, turned, opened the station house door, ushering out Julian and Feeney. "Now, I like you two boys. Come along with me to the Casa de Desierto. We'll have us some flapjacks and java before we head for Los Angeles. I'm on my way to meet my partner, Al Johnson and his lovely wife Mabel. They're staying at the Alexandria Hotel. They'll be coming out to the ranch next week. I'll be staying in one of those movie star suites they have at the Alexandria." Scotty led the two younger men along the gravel boarding area, walking eastward a few yards to the entrance to the Casa de Desierto.

The Casa de Desierto, a two-story structure with a front portico supported by thick granite columns, served as a fulcrum for the activities of a small but well appointed hotel. The three men entered through large wooden doors from the parlor of the Casa de Desierto into a ballroom and dining area. Scotty was greeted by a "Fred Harvey Girl" who escorted the threesome to a corner table with a north easterly view of the Mojave River. Harvey Girls, with friendly faces, served as waitresses at Fred Harvey Restaurants. These restaurants, located at principal stops along the main line of the Atchison, Topeka, and Santa Fe Railway, provided a place to rest and get good food at fair prices. Beginning in the 1880s, thousands of young women from throughout the East Coast and some foreign countries answered the call of adventure and headed West to work for the Fred Harvey Restaurants at locations beginning in Chicago and ending at the Los Angeles train station.

On a side rail about twenty yards from the dining area, a Pullman car, apparently private, was being watched over by two trainmen. A Harvey Girl waitress approached Scotty's table, placed tumblers of ice water on the white linen table cloth and offered the men menus. "Good morning, Mr. Scott."

"Good morning, Miss Epicure. How's the flapjacks?"

"As always, the best for you, Mr. Scott."

Scotty closed his menu, "Trust me boys. They have buttermilk or regular. I like the buttermilk. Three orders of buttermilk flapjacks and java. Is that OK with you boys?"

Feeney and Julian nodded agreement. The waitress jotted down the order on a Fred Harvey tablet, retrieved the menus and sped off to the kitchen. "So you heard about my Scott Special run in '05, Beeney?"

"Yes sir—I was in grade school and..."

"I paid the Santa Fe $5,500 cash to take me and my dear wife from Los Angeles to Chicago in not one minute less than 46 hours. We broke the record—44 hours and 54 minutes. Old Charles Losee was our last engineer...pulled us into Dearborn Station doing almost 59 miles an hour."

The waitress returned with three steaming mugs of coffee. "The cream and sugar are right there gentlemen." She pointed to a bowl of sugar cubes and a small pitcher of cream, "The pancakes will be here shortly."

Chauncey Julian surveyed the dining room. People he had seen on the train were having their breakfast along with neatly dressed couples and

< 11 >

businessmen who were guests at the hotel visible just outside the dining room window.

Bill Feeney snatched the lull in the conversation to try to direct Death Valley Scotty's conversation away from his railroad exploit. "Mr. Scott, you have been out here in the West for many years. What advice do you have for a couple of chaps like us to become successful in Los Angeles?"

Scotty seemed to reach back in time to provide the elixir for ambition to share with the two younger men.

"Partners, play the crowd. Forget them big shots in their mighty places—play for the numbers. When I joined up with Buffalo Bill's Wild West Show, it was ballyhoo that made us. Ballyhoo here in the U. S. of A. and ballyhoo all over Europe. When we came a ridin' into the arenas, I always had in the back of my mind to give a show for the big crowd, not for some governor or royalty and the like. When you boys get to Los Angeles, remember what old Scotty told you."

The waitress, in a deliberately showy style, served the stacks of pancakes. She placed a bowl of butter balls on the table along with three tin containers shaped and decorated like little log cabins.

On the side track outside the dining room, the clang, clang, clang of an engine bell caught the attention of the guests. A switch engine was coupling with the private Pullman car. On a path between the dining room and the hotel, walking as if in step with the engine bell, was a procession of several well-dressed men and a woman clothed in a shining white evening gown. Long white gloves covered her hands and forearms. She wore a white wide-brim hat. She was poised and urbane, giving no heed to her entourage or to the guests who peered out of the dining room windows.

"You know her, boys?" Death Valley Scotty chuckled. His two companions appeared awestruck at the unexpected apparition in the early morning at the Barstow Station. The woman first, then, the men entered the Pullman car. "That's Sister Aimee, boys. Aimee Semple McPherson. She can help save you two boys from all them sinners up there in Hollywood. She is building a temple right there between Hollywood and downtown Los Angeles. She came out to my Castle with her architect. The architect tells me he doesn't like Moorish. I told him I don't like Moorish either. He asked me for the name of my architect. I told him I don't have any architect. Sister Aimee there, she is my kind of lady—plays to the crowds. She told me she is gonna have her own radio station. You boys can learn from that lady preacher, especially you, Chancey, wantin' to be a big oil man. Play to the crowds like old Scotty and Sister Aimee there."

The whistle of the Los Angeles bound engine blew twice. The passengers in the Fred Harvey House sauntered toward their Pullman cars.

Death Valley Scotty tossed a $20 bill on the table. "This is on me, boys." Julian and Feeney followed Scotty to the boarding area. The switch engine backed the private Pullman car of Aimee Semple McPherson to be linked to the last car of the train, a United States mail car.

Death Valley Scotty grabbed the arms of Julian and Feeney, ushering them from their Pullman car toward the rear of the train. "You boys and I are going to have a little poker game between Barstow and Los Angeles.

< 12 >

They have a section all laid out for me in the rear of the baggage car called the Lounge Library."

At the rear end of the baggage car, Scotty, Feeney, and Julian seated themselves at a wooden card table . Scotty lit a cigar. "You boys want a smoke? Here, have one." Scotty offered Feeney and Julian cigars, but they both declined. As the train pulled out en route to the next stop in San Bernardino, Scotty called out to John Finlay seated at the opposite end of the Lounge Library. "Come over here, John, we need a fourth man for a poker game."

John Finlay, a tall well dressed man, had railroaded since 1879 and was the engineer on the Los Angeles to Barstow run of the Scott Special on July 9, 1905, when he averaged 84 miles per hour between Los Angeles and San Bernardino. Death Valley Scotty grasped John Finlay's outstretched hand, "Scotchman, I want you to meet my friends. This here is Beeney, and this is Chancey." Feeney and Julian shook hands. Scotty pulled a deck of cards from his coat pocket. "Sit down boys. Oh Beeney, would you bring us that set of chips over there on the bottom of the book case. Boys, the old Scotchman opened a liquor store in Barstow after he retired. Pro-he-bitchin closed him down. Doing any business Scotchman?" "Some Scotty. Selling wine to churches. I'm heading for the Cucamonga Winery today. An old German there sells altar wine to me wholesale."

The conductor opened the baggage car door and called out in a loud voice: "Next stop, San Bernardino! We arrive in two hours. Please remain on board. A ten minute stop only. Arrival time will be noon. John, I have you logged for a stop at the Cucamonga Station."

The foursome pursued their poker game in earnest, dissipating the rather monotonous journey through the Mojave Desert en route to the summit of the Cajon Pass, nearly 4,000 feet in elevation. The mountain ranges to the east and west of the pass were covered with snow above 5,000 feet. The down slope run into San Bernardino followed the edge of a riverbed hugging the western range of mountains until reaching the San Bernardino Valley. At the San Bernardino station mail sacks were exchanged from the mail car and a few passengers boarded the chair car.

The rear door of the baggage care swung open; a riflebearing, tall young man with his weapon ported blocked the passageway to the U. S. Mail Car. "Gentlemen, remain seated. I am Postal Inspector Madeira. This is Santa Fe Railroad payday. How's the game going?"

"Look at the chips, boy." Death Valley Scotty pointed toward the several tall stacks of poker chips near Chauncey Julian. "Chancey Julian here had a string of luck, Inspector. I'm Death Valley Scotty. The youngster here is Beeney, a paperboy. This old timer is John Finlay. You're unloading his pension money."

"Pleased to meet you gentlemen. Proceed with your game. We're ready to move on."

The poker game recessed at the Cucamonga Station where John Finlay exited for his wine business. The train arrived at the Los Angeles Station on time. Scotty hailed a taxi, as he had wired Al Johnson and Mabel to meet him in the lobby of the Alexandria Hotel. Scotty grabbed Julian and

< 13 >

Feeney by the shoulders. "Chancey look here. You want to be in the black gold business, get to the hoipolloi—the little guys who spend the bucks for gas and oil and the like. There ain't enough big shots to spend lots of money. Besides, they always expect Annie Oakleys. And you Beeney, when you get to Hollywood, look up old Wally Beery and tell him Scotty remembers the time we split a quart of Old Granddad out in the desert when he was workin' on one of his movies!"

Julian and Feeney waved farewell to Death Valley Scotty as he drove off in a taxi.

A uniformed band accompanied by hissing steam from the train engine and its clanging bell, struck up the tune, "Onward Christian Soldiers." The band was gathered on the station platform near the private Pullman car at the rear of the train. A crowd of well-wishers, including a fastidiously dressed gentleman carrying a large bouquet of red roses, was assembled near the Pullman car's door. Several press photographers, including a motion picture newsreel camera man, stood in a group facing loud speakers and a microphone which were located on a small portable stage brought out when celebrities arrived. After a second lusty rendition of "Onward Christian Soldiers," Aimee Semple McPherson emerged from the Pullman car, stepped down to the train station platform, and ascended the small stage. Her entourage of important looking men stood behind her. The man with the bouquet of roses raised his hand for silence. Feeney and Julian moved to the rear of the train and joined the crowd. Feeney noticed that the badges worn by the band members indicated that they were Los Angeles firemen. The man with the roses spoke into the microphone: "Welcome home, Sister Aimee! As Mayor of Los Angeles, it is an honor to present to you on behalf of the City Council and the people of Los Angeles, this token of our love and esteem."

The crowd applauded. Sister Aimee approached the microphone, "Thank you, Mayor Cryer. Our trip to the East Coast enabled us to complete the financing for the new temple. God was very generous to us. The bankers were very pleased with the presentation of our architect, Mr. Arnold who accompanied us to New York. We also discussed with the New York bankers our plans for our own radio station. The radio station will enable our preaching the gospel to reach even more souls here in Southern California. May the Lord bless all of you for being here! Thank you! Thank you, one and all!"

Sister Aimee was escorted to a nearby limousine by two uniformed policemen. Four motorcycle policemen with red lights gleaming and sirens screaming proceeded out of the parking area towards Sunset Boulevard while the Fireman's band played, "The Battle Hymn of the Republic."

As the crowd disbursed, the Mayor and several councilmen entered two police cars assigned to drive the officials back to City Hall.

Bill Feeney shook hands with Chauncey Julian. "Well C.C., I think we had better part ways. I really wish you well in your new venture in Santa Fe Springs. Here's my phone number at the Wilcox Hotel. Give me a call when you have a chance."

"I'll give you a call as soon as I get settled, Bill."

< 14 >

The two men picked up their suitcases from the baggage area. Bill Feeney hailed a cab and was driven off to his hotel in Hollywood.

Chauncey Julian decided to walk to the Plaza. He seated himself on a bench in the shade of a large Magnolia tree. Across Main Street, Julian watched several children playing in an Indian graveyard to the south of the old Plaza church. The aroma of spicy food from the Olvera Street outdoor kitchens reminded Chauncey Julian that it was past lunch time. The clatter of a passing trolley car heading south on Main Street toward downtown merged with the shriek of the siren of a large red fire truck which was rushing toward Macey Street. He was alone with his dreams in the City of the Angels where men and women were grasping for and frequently obtaining lots of money, fancy clothes, and big automobiles to a degree unmatched anywhere else in the United States. Chauncey Julian removed a small notebook from his pocket and wrote the following equations. Death Valley Scotty plus the crowds = gold. Sister Aimee plus the crowds = the temple. C.C. Julian plus the crowds = money. Julian folded the little memo, placed it in his wallet, stood up, picked up his suitcases, and instead of walking, hailed a taxicab. "Driver, take me to the Lankershim Hotel."

CHAPTER 2

Bill Feeney's cab turned south off Hollywood Boulevard on Wilcox and after half a block, the driver made a U-turn stopping in front of the Wilcox Hotel. As the cab driver removed Bill's luggage from the cab's trunk, Feeney remarked, "I'm a bit hungry after the trip from Barstow. Can you recommend a good restaurant within walking distance?"

"Yeah, up on Hollywood Boulevard. It's called Musso & Frank's Grill. For lunch, I like their onion soup and French bread. Try them for breakfast. Best flannel cakes in the world!"

"Thanks, I'll take your advice—keep the change!"

Bill Feeney approached his new lodging. During the interview with the Citizen News personnel man on the campus in New York he inquired about lodging in Hollywood close to the newspaper and was referred to the Wilcox Hotel. While he was attending Columbia University, he had signed a rental agreement for three months and advanced a month's rent by mail. The rent money was a residue from a trust account which Bill Feeney's father had opened for him. The elder Feeney, the publisher of an upstate New York daily, assumed the burden of supporting his son up to the time of his employment on the West Coast.

In a courtyard before the entrance to the hotel, Bill noticed a fishpond built like a water well with a spout of water above the pond flowing into a circular basin, causing a miniature waterfall effect. Goldfish were puckering at the surface of the pond between the water lilies. Not something for New York in February, he mused. Bill was greeted at the registry desk, signed in, given two sets of keys, and shown the way to the second floor room as the hotel had no bellboy or elevator. His room had a view of Wilcox and

< 15 >

the Hollywood Citizen News building across the street. By stretching a bit outside his window, Bill could see Hollywood Boulevard half a block to the North. The room was immaculate with a queen-size bed, small kitchenette, and dining room. It included a private bath and shower. Bill plopped on the bed, staring at the ceiling beginning to feel a bit alone in a city which he hoped would be the successful maiden voyage of his journalism career, and perhaps would serve as a matchmaker for the woman of his desire.

"My first phone call!"

The telephone on the shelf between the bedroom and the kitchenette rang sharply jolting Feeney to his feet. He had not noticed the phone in his quick survey.

"Hello, this is Bill Feeney."

"Welcome to Hollywood. This is Hank Bond over at the Citizen News. I'm the city editor and I would like to set up a meeting for Monday morning. How about 8 am.?"

"That's fine, Mr. Bond, I'll see you then."

"Listen Bill, one thing Art did not discuss with you when you were hired, was transportation. We don't provide a car. You can get around this town on red cars and trolleys, but I think you may want to consider purchasing your own automobile. I understood by your wire that you were arriving on the Santa Fe today, so assumed that you didn't have a car."

"That's correct, Mr. Bond." "Call me Hank, Bill."

"O.K., Hank. Can you give me a lead on a good used car?"

"Try the used car lot of the Cadillac dealer, Hillcrest Motors, on the northeast corner of Hollywood Boulevard and Orange. It's a bit of a walk, about five blocks west from where we are. When the actors and actresses out here make the big time, they trade in their cheaper cars for Cadillacs or Pierce Arrows and the like. A fellow can get a good buy there."

"Thanks, Hank, I'll see you Monday." "By the way, if you do buy a car, you can park it in the newspaper's parking lot. I've given the watchman your name and I.D. number and he'll direct you to your parking space. I'll see you Monday."

"Bye, Hank."

Bill Feeney loosened his watch from his vest pocket, flipped open the gold cover and noticed that it was well past lunchtime. He decided to dine early and retire early. He showered, shaved and dressed for dinner. Since he did not know his way around town, the suggestion of Musso & Frank's Grill seemed to be the best bet. Bill left his hotel and walked toward Hollywood Boulevard. A few yards from the hotel, he discovered an elongated newsstand parallel to the sidewalk, constructed along a wall of the building on the corner of Wilcox and Hollywood Boulevard. The newsstand was sheltered by a green awning. A twinge of nostalgia as he noted the prominently displayed New York Times was displaced quickly as he walked briskly down brightly lighted Hollywood Boulevard. After dinner, Bill thought he might check out the new Grauman's Egyptian Theater which recently hosted the premiere of Robin Hood. Bill crossed Hollywood Boulevard at Cherokee and slipped through the double doors of Musso & Frank's Grill. The dark knotty pine walls, copper light wells, wooden

< 16 >

panelled dining booths, and long dining counter gave the impression more of a hunting lodge than an eatery in glamourous Hollywood. Although early in the evening, the hostess suggested a seat at the counter rather than a booth. Bill found a spot halfway down the counter directly in front of a trio of chefs, dressed in appropriate white caps, white waistcoats, and aprons who were busy broiling steaks and chops on an enormous grill constructed from stove top to ceiling with shiny red bricks. A busboy with white shirt, black bow tie, and green vest simultaneously placed a plate with French bread wrapped in a white napkin, a small plate with butter squares, and a goblet of ice water near his table setting. A red-jacketed waiter supplied a menu.

"Good evening, sir." The waiter leaned over the counter and whispered, "Would you like something from the bar?"

Bill had not noticed any evidence of a bar on entering the restaurant, but agreed to the idea and whispered, "Rye and soda."

The waiter whispered, "All we have is Old Overholt, sir, will that do? Prohibition, you know."

"Yes, Sam." Bill noticed the brass name tag on his lapel. The waiter disappeared around a wall to Bill's left, so Bill assumed the bar was behind the grill in a closed section of the dining room. Bill decided on a New York steak, medium rare, baked potato, and a salad with the house dressing. Sam, the waiter, brought Bill his drink, took his order, and relayed the order to one of the chefs. As he sipped his drink, admiring the culinary expertise of the chefs directly in front of him, Bill was slightly jostled as a woman squeezed into the unoccupied swivel chair to his left.

"Pardon me, they seem to place these chairs so only midgets may enter easily."

As she slid into the chair, the woman tossed her head so that her waist-length golden hair fell comfortably behind the back of the swivel chair.

"Good evening, Miss Kameal."

Sam, with energy not previously evidenced, produced a menu seconds before the busboy arrived with the bread, butter, and ice water.

Sam whispered, "Would you care for anything from the bar, Miss Kameal?"

"The usual, Sam. They let us off early tonight." "Right away, Miss Kameal."

Sam again disappeared behind the wall.

"You seem to be well known here, Miss Kameal. Perhaps Sam will call me Mr. Feeney in a week or so."

"Don't flatter yourself, Feeney!"

The young woman turned and looked straight into Bill Feeney's eyes holding contact as though in a contest to describe their precise color.

"Feeney, you don't look like an actor. What brings you to this town?"

"I'm a reporter. I just got hired by the Citizen News. I start work Monday."

Sam returned, placing with a flourish of the napkin draped over his arm, a glass of amber colored wine near Miss Kameal's folded menu.

"I will have the lamb chops on the dinner menu and a bowl of onion

< 17 >

soup a la carte. Sam, meet Mr. Feeney, a newspaperman, a new arrival."

The waiter smiled expansively, jotting down Miss Kameal's order for the chef. "I hope you will dine with us often, Mr. Feeney!"

Turning to the chef again, Sam called for an order of lamb chops.

"Here's to your success in Hollywood." Miss Kameal raised her wine glass to Bill Feeney in a gesture of a toast. "Salud!"

"Cheers, and thanks for the unexpected hospitality to a newcomer. By the way, call me Bill, Miss Kameal, and you are...?

"Kathryn is my real name. My stage name is Kay. Kay Kameal."

"So, you're an actress?"

"I am indeed a real live Hollywood actress, presently soon to be between things, as they say. Have you seen Robin Hood?"

"Not yet."

"It's showing second run around town. In the scene when Mary Pickford, the Queen, sends Douglas Fairbanks on a mission as a knight, I am one of the ladies in waiting. Fourth from the left."

"I am impressed. This is my first interview with a celebrity. I should have carried my notebook."

"Don't try to be funny, Bill, you're no comedian."

"Sorry, no offense, I really am impressed. I've never spoken to anyone connected with motion pictures, and I really hope to do some stories about the Hollywood community although I was hired to cover City Hall and politics downtown. I don't know very much about politics but I did see the mayor of Los Angeles and some councilmen at the train station, welcoming home Aimee Semple McPherson."

Sam served the soup and salads and replenished the French bread.

"Tomorrow, I'm going to look for a used car. My editor, Hank, tells me I'll need one to get around this place. It seems so spread out."

"Yes, you will need one. I have a Model T that my parents bought. It's parked behind the restaurant. It cost $350 new. I drive from my apartment to the studio every day. Cabs are too expensive and they're hard to find."

"Where is the studio?"

"I work at the Pickford/Fairbanks Studio on the corner of Santa Monica and La Brea, about three miles from my apartment. I'm staying at the Garden of Allah on Sunset Boulevard."

The busboy removed the soup and salad dishes and Sam picked up the steak and chops from the sideboard near the grill, scooped up some peas from a vegetable hotplate, added a mound of mashed potatoes next to the chops, took a baked potato from an oven to the right of the grill, and turned, placing the steaming plates before the couple while the busboy placed bottles of Lea & Perrins and A-I Sauce nearby.

"Care for any coffee or tea, Miss Kameal and Mr. Feeney?"

"Not just now, thanks, Sam."

Kathryn responded for both herself and Bill Feeney, then nudged Bill Feeney."

"Mr. Feeney, are you Irish?"

"SWell, yes."

< 18 >

"Are you Catholic?"

"Yes—I mean, well—yes, but—"

"This is Friday, you're not supposed to be eating meat, are you?"

"You caught me. This is definitely a situation I will need to bring up this Easter."

For dessert, Sam suggested the cheesecake. Two orders arrived with coffee served in small metal pots. Bill Feeney insisted on picking up both tabs, leaving an ample tip for Sam and the Busboy.

"Look, Kathryn, I'm going to buy a car tomorrow. May I stop by your apartment Sunday morning and take you for a spin?"

"If you like. About eleven. The Garden of Allah is on the south side of Sunset, 8259. My apartment is the last on the left. I'll see you Sunday then, and thanks very much for the dinner. Good luck in your career in Hollywood."

Kathryn Kameal spun her head quickly, brushing Bill Feeney's shoulder with her golden hair. She waved goodbye to Sam and left Musso & Frank Grill through the rear door. Bill Feeney returned to Hollywood Boulevard. His desire to visit the new Grauman's Egyptian Theater was dampened by the unexpected and welcome encounter with Kathryn Kameal. He returned to the Wilcox Hotel. Bill asked the night manager for any messages.

"Yes sir, you had two phone calls. One is from your mother. She asked to have you call her tomorrow collect. The other is from 'C.C.' He left his address and phone number at the Lankershim Hotel."

"Thanks."

Bill Feeney took the two scraps of paper, climbed the stairs to his room and retired, sensing that he had become, after crossing the border from Arizona to California, a participant in a way of life not completely expected, giving him some trepidation as he seemed to have lost a degree of control— not necessarily of his career, but more accurately of his destiny.

After a deep sleep, Bill Feeney, an early riser, walked to the newsstand, bought a copy of the morning paper and went into a donut shop on Hollywood Boulevard, "Two powdered donuts and coffee." Bill unfolded the paper and read the headline.

WHO WILL CONTROL UNION OIL COMPANY?
PROXY BATTLE INTENSIFIES.

Uninterested, Bill Feeney turned pages hoping for some news from home. He returned to his hotel room and placed a call.

"You have a collect call from William Feeney, will you accept the charges?"

"Yes, Operator."

"Hi, Mom. I'm fine, I arrived yesterday. I took a cab to my hotel. Do you have a pencil handy?"

"Just a minute—go ahead, Bill."

"Here's my new address, 1530 Wilcox Avenue, Room 202, Hollywood, California. How is Dad? Is he working today?"

< 19 >

"Yes, Bill, and he gives you his love."

"In about an hour, I'm going to look for a used car. A car is a necessity out here."

"Do you have enough money, honey?"

"Yes, I have plenty of money from the trust account and I'll get my first paycheck next month. Oh, by the way, I met a Hollywood actress at dinner last night."

"Bill, we don't want you running around with any floozy actresses."

"Now, mother, trust my judgment. She's really very nice. Have you seen the movie *Robin Hood*?"

"No, Bill, I haven't."

"Go see it. When the scene arrives with the Queen sending Robin Hood off to do her bidding, my friend is a maid-in-waiting, fourth from the left."

"Now, Bill, you're not getting serious after one encounter?"

"No, I'm not serious, I've only been here two days."

"Please be careful, Bill."

"Yes, mother, I'll be careful. Say 'hello' to Dad. I'll write soon. I love you."

"Goodbye, Bill. I love you."

"I love you, too, Mom. Bye."

Bill Feeney approached the line up of used cars. In the front line were several used Cadillacs and an Auburn Coupe. In the second row were a several Model Ts, three Pierce Arrows, a Cleveland Six, a Hupmobile and two Dodge Brothers sedans. A salesman emerged from a small office in the middle of the lot. He was neatly dressd—dark suit, white shirt, and a somber colored necktie.

"Good morning, sir, may I help you? My name is Walter Hutson." The salesman handed Bill Feeney a business card.

Bill Feeney examined the card. It had a Cadillac emblem enbossed on the left corner, presumably to help establish the credentials of the used car salesman.

"Thanks, Walter. I'm looking for a transportation car, mainly for work. I will be doing a lot of driving on the job so I want something durable, but not too flashy. On weekends I will be using it for pleasure. You know, taking my girlfriend out for rides and such."

"Your name, sir?"

"Oh, I'm sorry, I'm Bill Feeney, just in from New York. I'll be starting work at the Citizen News Monday."

"Pleased to meet you, Bill." The two men shook hands.

"Bill, we just took in this Auburn parked next to the office." The two men walked several yards over to the used car office.

"This is a Six Supreme, it's also called the Auburn Beauty Six. It's two years old, has low mileage and is a one-owner car. A screen writer who was signed to a long-term contract over at the Pickford/Fairbanks studio moved up to a Cadillac convertible. Take a look at the genuine leather upholstery, both front and rear. This particular model has a permanent top,

< 20 >

however, there's also a model with a folding top. It has a 63 horsepower engine and will move from one mile an hour to 50 miles an hour in eleven seconds. Would you like to take it for a test drive?"

"Sure, let's go." "Bill, do you have a drivers license?"

"Yes, I have my New York license."

Bill with the salesman seated at his side, drove east on Hollywood Boulevard, then north on Highland, toward the Hollywood Hills.

"It feels pretty solid, what's it worth?"

"Seven hundred and fifty dollars."

As the car gained speed passing the Hollywood Bowl, Bill decided. "Very well, I think I'll take it."

"Stop for a minute, Bill, and let me drive. I want you to sit in the rear seat to get a real feel of comfort. This car is manufactured in Auburn, Indiana. They began making Auburns in 1903. A few years ago, the company was purchased by the chewing gum manufacturer, William Wrigley, Jr., and a group of Chicago businessmen. They are careful to make parts available all over the United States. Jack Urban, the sales manager likes this car and maintains a stock of parts available at all times. By the way, Jack Urban also told me he went all over this car and it is in A-I condition."

Walter Hutson returned the Auburn to the parking slot next to the used car office. "Bill, come on into the office and we'll do the paperwork,"

An hour later, Bill Feeney drove his Auburn to the Hollywood Market, did his grocery shopping, drove over to Melrose to purchase some sports clothes, and returned to his hotel, awaiting Sunday morning; and Kathryn Kameal.

The Sunday morning drive on Sunset Boulevard proved to be uncrowded. Hollywood folks, generally early risers for make-up and set preparation; on weekends, were party-goers by night and recluses before noontime.

Bill Feeney drove westbound on Sunset picking up addresses visually until he reached the 8000 block of Sunset Boulevard. "8259" was numbered on the front of an apartment complex on the south side of Sunset near Crescent Heights Boulevard. Making a left turn off Sunset on Crescent Heights, Bill pulled his Auburn to a stop, parking parallel to the curb in front of a newly constructed set of flats, small, Spanish style; each with a red tile roof, and each with a single car garage. Bill noted a "For Rent" sign with a phone number and jotted it down for future reference as a possibility for a permanent residence. Jaywalking across Crescent Heights, Bill returned to Sunset and entered the front patio of the Garden of Allah apartments. As he walked toward the last apartment on the left, Bill found his way through the debris of someone's Saturday night garden party. Tables were cluttered with empty unlabeled wine bottles. Paper plates with half-eaten food were strewn about. Bill knocked lightly. Kathryn Kameal opened the thick wooden door to her apartment The bright sun piercing through the palm trees in the apartment complex courtyard embellished Kathryn's long flaxen hair, held back from her forehead by a dark blue fillet. Her one piece light blue and

< 21 >

white wool dress was about knee length. She wore a blue sweater as the weather, though sunny, was brisk.

"Good morning, Bill. What make of car did you buy?"

"Come along, I'll show you."

"I think I got a good deal. Walter Hutson, the salesman, told me that it was a trade-in from a screenwriter who made it big. By the way, who threw the party last night?

"The apartment house owner is a former Hungarian actress. She invited a group of her friends. They were all speaking Hungarian. Despite the mess in the courtyard, the party did not get out of hand. The apartment owner made a bundle of money in films. She has the idea that talking pictures are coming. She knows her accent would ruin her, so she invested in this apartment building. She lives here and supports herself through the rentals."

Walking around the corner, Bill pointed to the Auburn.

"That's it, hop in. Where to?"

"How about lunch at the Ambassador Hotel over on Wilshire Boulevard?"

"How do I get there?"

"Take Sunset to La Brea, turn right to Wilshire, make a left turn, it's about three or four miles."

"That's it, Bill, make a right turn at the next signal. The hotel is set back in a private park. The parking lot is in the rear." Bill pulled his car into a distant parking stall.

"I don't want to get any scratches on this car if I can help it."

They strolled through the park-like gardens, past the swimming pools without bathers despite the warming trend of the weather.

"You lead the way, Kathryn You seem to know your way around."

"Follow me. That's the Coconut Grove. When Hollywood producers want to promote a new star, they put on a big party there. We'll go down to the coffee shop below."

The hostess at the coffee shop seated the couple in a corner table with a view of the west garden area. As Bill and Kathryn were handed their menus, a commotion at the door drew the attention of those already seated. A group of eight to ten men and women, conversing loudly, were led to a large table with a "reserved" sign. Most of the men wore dark hats which they hung on a hat rack near the entrance to the Coffee Shop. Members of the group seemed to defer to a slender, silver-haired distinguished looking man and his wife who dressed conservatively and expensively in Eastern designer's winter clothing.

"Kathryn, do you know that tall man seated at the head of the table?"

"No, I've never seen him before. He looks wealthy and powerful."

"Wealthy he is, powerful he was. That's former Secretary of the Treasury, William Gibbs McAdoo, and his wife. I read in the New York Times that they were moving to Southern California. His wife is the daughter of Woodrow Wilson. Do you remember the 1920 Democratic Convention? He was a candidate for President but didn't make it."

"Vaguely, I don't pay that much attention to politics Let's have

< 22 >

lunch and drive down to Santa Monica for some sea breeze and less hot air from politicians."

Bil parked the Auburn on the Palisades above the beach in Santa Monica.

"Bill, let's take a walk."

"Great, I brought my camera, may I take your picture?"
"Sure, or my agent can send you a glossy."

"You can do that also, I would like to have one. What I'd like to have is a snapshot to send to my mother. A glossy might overwhelm her, especially if it's a scene from *Robin Hood*."

"That's nothing, you should have seen the glossies for *The Affairs of Anatol*, or did you see the picture?"

"No, I must have passed over that one."

"When you send your mother the snapshot, it may be reassuring to her that I have signed with Cecil B DeMille for, *The Ten Commandments*. That should calm any anxiety about her son out here in this wicked place."

"Please, Kathryn, I'm no Mama's boy. It's just that I can't introduce you to my parents when we're so far away."

"I'm far away from my parents, also. I was raised in Chicago. Father is an attorney. He didn't think there was much future for a woman practicing law. I had a flair for drama and choir in high school, so my father paid my tuition, room, and board at USC. I majored in Liberal Arts; studied Drama and Chorus as a minor. I joined the Hollywood Community Chorus and sang last Easter at the Hollywood Bowl's first Easter Sunrise Service. I graduated in June and Mrs. Artie Carter, the principal promoter of the Hollywood Bowl and patron of the Hollywood Chorus called the people at the Fairbanks/Pickford studio. I've been working steadily ever since. I will be singing at the Bowl this Easter. Will you come with me?"

"I'm flattered, certainly. I like you very much. During college, I didn't date very much. I was kind of a bookworm and kept plugging away to finish school as quickly as I could. I really hope our careers have led us to one another. I hope we'll see a lot of each other."

"I hope so too, Bill."

Kathryn Kameal placed both arms on Bills' shoulders and kissed him quickly. Bill grasped her waist, pulled her gently to him, looked squarely into her eyes, "Kathryn; can you tell that I'm not just another actor?"

"Yes, I can, Bill."

CHAPTER 3

"Good morning, I'm William Feeney reporting for work. I'm to see Hank Bond at 8 a.m."

"Good morning, Mr. Feeney, I'm Maggie McCloud — just call me Maggie. Mr. Bond left this press card for you. I need to check your identity."

< 23 >

Bill removed his New York drivers license from his wallet. She verified the date of birth and returned both the press card and drivers license.

"I'll ring Mr. Bond, just a moment. Mr. Bond, Mr. Feeney is here. Mr. Feeney, just go through that door to the left, up one flight of stairs. Mr. Bond's office is on the right as you enter the newsroom."

"Thanks, Maggie."

Bill entered the small but well organized newsroom. Half a dozen reporters were busy at desks, each with his own typewriter and telephone. Several of the staff were lounging in what appeared to be a ready room or conference room, drinking coffee, and chatting. There was a wood and glass enclosure with "City Editor" etched on the open door, Hank Bond smiling, shook hands with Bill Feeney. Bond, a gray-haired, rather heavy set, individual, mid-forties to early fifties, wore a green visor on his forehead. A black cord supported reading glasses dangling on his chest. An unlit pipe rested in an ashtray on his roll-top desk.

"Sit down, Bill, happy to have you on board."

Hank Bond swiveled in his chair, picked up the phone and dialed.

"Maggie, will you phone the photo lab and have Raul Dominguez come to my office? Thanks."

"Bill, Raul is a senior at the University of California over on Vermont. He works for us part time. His father is a native Angeleno. Raul knows his way around this town and through his father's connections he knows at least by name and sight, all the old timers and people who have power here. Even if you don't need a photographer, he can be a real asset. He can help you on your assignments and sometimes go with you." There was a knock at the City Editor's door.

"Raul, come in here. I want you to meet Bill Feeney, the new reporter I told you about. He's a New Yorker. The Personnel people tell me he's adaptable and will quickly get straight as a string about things in Los Angeles."

Raul Dominguez greeted Bill Feeney with a firm handshake. Raul was an inch or so taller than Bill, with athletic features, a swarthy complexion, broad shoulders, and a slender waist. He wore gray trousers, a blue shirt, and a slip-over sweater—a college-type.

"Now you two sit down and we'll discuss your assignments for the next week or so. Raul, I know how you work and I think you'll be a big help to Feeney here. Bill, most of Raul's classes are late afternoon or early evening so he'll be available on a daily basis when you need him. I'm taking him off sports so his time won't be tied up. Bill, our needs here were outlined to you during the interviews at the university. You were hired to cover City Hall with an emphasis on what we want not what the Los Angeles downtown people want us to hear. Our readers want to know how City Hall will benefit folks in Hollywood, Beverly Hills, and the San Fernando Valley. For your information, Bill, Hollywood was annexed to Los Angeles in 1910. We needed their water to survive so we had to join them. However, this is still Hollywood. We cherish their water and use their police and fire departments. They, on the other hand, use us because of the world-wide attention given to Hollywood through our business - movies. Basically, the down-

< 24 >

town power boys give us little in return. So when you cover City Hall, look for angles of interest for our readers. Another thing, apart from dabbling in real estate, many of our readers are stockholders of various oil companies. I want you to cover the oil companies as best you can."

"I met an oil man on the train. I spoke to him at the Fred Harvey House in Barstow. A fellow named Chauncey Julian."

"I've never heard of him but they come here in droves. Wells everywhere. They all think they're going to become wealthy fast. I picked up this book, "Business of Oil Production". You can keep it, Bill. I want you to read it so when you ask the oil people questions they will make sense. That's about it. Cover City Hall and the oil business. You make your own schedule. Your deadline is 1 pm. for the afternoon edition. You'll be paid once a month on the fifteenth. Your check will be ready by noon. Maggie at the switchboard is our paymaster. Any questions?"

"Not really. What is our first assignment?"

"Thanks for asking. At 10:30 in the director's room of Security Trust and Savings Bank downtown, Mayor Cryer will hold a press meeting. Security Bank two years ago merged with the Hollywood National Bank and the Citizens Savings Bank of Hollywood. The downtown people see Hollywood as a part of their turf. They're allowing the mayor to use their director's room as the bank is opening branches all over Los Angeles. There's a rumor that the City may sue Union Oil Company. The rumor has it that the City may try to have its own harbor and somehow the Union Oil Company is in their way. See if you can get a line on that story. Raul, show Bill his desk. Good luck, and see you later."

Bill and Raul entered the newsroom. Bill's desk was near a window overlooking the newspaper's parking lot on Wilcox.

"Raul, see that Auburn? Get your camera. I'll meet you at the car in ten minutes and we'll head downtown."

"Right, Bill, see you in ten."

The directors' room at the main office of the Security Bank at 5th and Spring was as stuffy and cold as that of any East Coast banking establishment. The Board of Directors' gleaming mahogany table, with ornately hand carved wooden supports at each end was surrounded by ten wooden chairs richly upholstered in red leather. Seated were city officials with Mayor Cryer at the head of the table near a marble fireplace. Two executive desks and chairs were located to the left and right of the table. Inside the double entry doors, a dozen or so reporters sat on wooden folding chairs arranged in two rows. The major portion of the parquetry floor was covered by a large Oriental rug. On the paneled wall to the left was a large portrait of the man who was seated at the opposite end of the directors' table from the Mayor. Guarding the double doors was an elderly black man in a dark blue uniform, a bank security officer. Raul Rominguez produced his press card and Bill Feeney followed the cue.

"Good morning, Mr. Shores. Bill Feeney is our new reporter at the Citizen News."

"Good morning, sir."

< 25 >

Shores handed Bill his press card after examining it briefly.

"How's your father, Raul?"

"He's fine. I'll tell him you inquired. Is Mrs. Shores well?"

"Very well, Raul. We are both up there in age like your father."

"You two had better be seated as this meeting will begin in a moment or so. I'm very pleased to meet you, Mr. Feeney.

"Bill, Mr. Shores has been an employee of the bank for over thirty years. His family members were among the founders of this city."

A large wall clock struck 10:30 am. The gray haired, mustachioed, elegantly dressed man—the living image of the portrait on the wall—rose. The chatter hushed.

"Gentlemen, I am Joseph Sartori, president of the bank. I wish to welcome Mayor Cryer and the council members and members of the press. The Security Bank welcomes opportunities to make our facility available for civic affairs. I trust that the Mayor's words will be good tidings for our city and neighboring communities. Mr. Mayor, welcome! I bid you all good day."

Mr. Shores moved to Mr. Sartori's side and the two men exited the chambers through the double doors. All eyes were turned toward Mayor Cryer who shuffled some papers, stood, and began the conference.

"Gentlemen, my message for the people of Los Angeles this year is the need for a new City Hall. Presently, eighty percent of the city's offices are in rented buildings. The rental cost is enormous. So, with the concurrence of the majority of the City Council, we will be encouraging our constituents to help us build a new City Hall. This will help us keep pace with the growth of this city. For example, the value of building permits in 1920 was $82 million. Last year, it was $121 million. We are quickly approaching New York and Chicago in the area of new building permits. This year we expect 70 new tracts of homes on 70,000 approved lots. Our population is growing at the rate of about 100,000 persons per year.

In view of this growth, I am pleased to announce that the city of Los Angeles for $13 and a half million has purchased the Edison Company Distributing System. This purchase was part of the plan conceived by former Mayor Fred Eaton, who pioneered the Los Angeles-Owens River Project.

Lastly, I wish to announce that our city is close to having a new charter, the first since 1889. Our city employees now exceed 5,000 again showing our need for a new City Hall. Thank you!"

Mayor Cryer seated, placed his working papers to the side. He folded his hands, stared at the reporters, and asked, "Any questions, gentlemen?"

"Mr. Mayor, Sam Salsbury, from the Times. Can you give us any cost estimate on the City Hall project?"

"Not yet, Sam. We are just looking at preliminary plans from several architects. We are looking at architects' renderings for both architectural and practical aspects of the project. Also, the seismic issue is very important for us."

"Mr. Mayor, Griffin, from the Examiner. There's a rumor that you are going to Hawaii. Is that true and is it for a vacation or for city business?"

< 26 >

"Clay, indeed I am. The date is not yet set, I will be a guest on the new liner City of Los Angeles on her maiden trip to Honolulu. I hope to encourage more commercial activity between the islands and the San Pedro harbor."

"Mr. Mayor, Cliff Brand, from the Long Beach Press Telegram. There are rumors around our City Hall that the City of Los Angeles is about ready to annex a strip of land down to and including San Pedro to try to force Union Oil Company to relinquish its lease portion in the San Pedro harbor. Is there any truth to that?"

"Sorry, Mr. Brand, I can't comment on that at this time. I will say this. Due to the strong probability of more trade through the canal completed in Panama, the city of Los Angeles is certainly interested in exploring the possibility of having its own salt water port."

"Mr. Mayor, Bill Feeney from the Hollywood Citizen News."

"You're a new face, Mr. Feeney."

"Yes sir. Another question about Union Oil Company. Do you believe Union Oil will be able to resist the takeover move on the part of Royal Dutch Shell?"

"Well, perhaps you should ask Lyman Stewart, their board chairman. But I will say this, if Union Oil should fall to foreign interests, it would be a disaster for our community."

"Thank you, Mr. Mayor."

"That's about it, gentlemen."

The mayor walked toward the double doors followed by the other politicians. He paused near Bill Feeney, placing his hand on Feeney's shoulder, "Mr. Feeney, I suggest you attend the Union Oil Company's board meeting March first for some real Los Angeles fireworks."

"Thanks for the advice, Mr. Mayor."

Bill Feeney and Raul Dominguez returned to Bill's car parked in a parking lot on Spring Street.

"Did you get a shot of the mayor standing? He's more impressive standing rather than seated in that banker's chair."

Bill filed his story by the deadline and walked to his hotel, pausing at the manager's cubicle. "Any messages, Fred?"

"Yes, Mr. Feeney. A Mr. Julian called and left his phone number, also a Miss Kameal called. She said she would be on the set late this evening and will call you around 6 pm."

"Thanks, Fred." Bill Feeney opened his window, took off his coat, tossed it on a chair, and called Chauncey Julian.

"C.C., how are things going? This is Bill Feeney."

"Bill, I think things are going to come together this go around. Listen, I have some free time Friday, I'm going to be fitted for some new suits. Could you meet me in the lobby of the Alexandria Hotel and have a drink or two?"

"Sure, C.C., how about 8 p.m.?"

"Fine."

"By the way, C.C., I would like to introduce you to a wonderful

< 27 >

woman. She's an actress, but very unaffected. Her name is Kathryn Kameal. I met her casually at a local restaurant."

"I would be pleased, Bill. I'll see you both there at 8. Bye."

Friday evening at the Alexandria Hotel was a time for the business and banking elite of Los Angeles to loosen up after the work week. Chauncey Julian was dressed impeccably. He had a rosebud in his lapel and was sporting a gold watch bob and chain across his vest. He carried a leather briefcase. He seated himself on a velvet settee guarded by a wooden rail near the Palm Court Ballroom. A player piano across the lobby was hammering out an unfamiliar tune. Julian motioned to an idle bellboy standing near the cashier's desk.

"Young man, I have a task for you. Here's five bucks. This is my business card. In half an hour, page me, first in the lobby, then in the coffee shop, and finally, in the Charley O's Bar. Got it?"

"Yes, sir."

"Bill Feeney." Chauncey Julian moved out of the gated area and shook hands with Bill Feeney, turning politely and extending his hand to Kathryn Kameal, who received his firm grip and unexpectedly a bow and a kiss on her hand. Kathryn was wearing a black decollete dress. A black fillet on her forehead enhanced her pearl earrings.

"Miss Kameal, I'm very pleased to meet you. I'm happy you could come along with Bill to meet me for a drink. Shall we?"

The trio entered the Charley O's Bar off the main lobby and in a darkened atmosphere, found a corner table.

"Did you notice the motto on the door of the bar, Bill?"

"No, C.C., what is it?"

"Where the world's problems are solved daily. I like that."

A waiter approached. "Good evening. Gentlemen and Madam?"

To Chauncey Julian who seemed to be in charge of things, "Yes, sir?"

"Three set ups."

"C.C., before I forget it, I brought you a book my editor made me study. Here." Bill Feeney drew a small black covered book from his coat pocket and handed it to Chauncey Julian, 'The Business of Oil production', by Roswell Johnson."

"Thanks, Bill, that's exactly what I'm going into in a big way." Julian opened his briefcase and deposited the black book.

The waiter brought three glasses, an ice bucket, and a seltzer bottle from a nearby bar, placing white doilies on the table.

Chauncey Julian poured double shots from a flask into each glass. "Ice and soda for both of you?" The couple nodded agreement.

"Here's to your success, C.C."

"Cheers," chirped Kathryn.

"Well, Bill and Kathryn, I'll tell you this, oil business around here is tough. The big boys control the bankers and grab off oil leases faster than jack rabbits. I call them, 'the respectables.' They all know one another and

< 28 >

help each other sell the stock that keeps their operations afloat. A lease I had last year near Huntington Beach came in dry as I told you . I'm using my savings to do some drilling at a lease I have in Santa Fe Springs. By the way, remember I told you I had contacted some boys over in Arizona just before I met you?"

"Yes, I recall."

"I received a letter from them yesterday. They're going to cut me in on a lead mining operation near Bisbee."

"Sounds interesting. Why lead, not gold or silver?"

"They tell me there's definitely going to be a world shortage of lead and we're going to be in on the ground floor."

"I'd check into that more if I were you C.C."

Kathryn broke into the conversation, "C.C., you're fascinating. Is there a lady in your plans?"

"Yes, Kathryn. I have a wife, Mary Olive. We were married in my native Canada on December 1, 1909. As soon as I get things off the ground here, my plans are for her and my daughters, Lois and Frances to come from Canada to live with me while I do business here."

"C.C., I've been assigned to cover the oil industry. My editor claims many of our readers are investors in oil stock and he wants me to keep them informed so they won't be led astray. I'm going to the Union Oil Company board meeting next week."

"Bill, when you get to that meeting, keep in mind what caused their problem with Royal Dutch Shell. The respectable folks at Union Oil, Milton and Lyman Stewart basically want to promote a Bible Institute here in Los Angeles to save the young folks and they also want to send Bibles to China. So what do they do to raise cash? They sold off big blocks of Union Oil stock to finance the purchase of Bibles and the Bible Institute. They had noble motives, but Royal Dutch Shell was not so noble. Royal Dutch Shell bought the stock and started Union of Delaware and are well on their way to take over Union Oil of California. Union Oil stock is selling for $160 a share. There's no chance for the little guys in Los Angeles to make money in Union Oil stock. I'm going to give the folks out here a real opportunity to ride with me to the top." His eyes glistening, Julian opened his briefcase, pulling out a lease agreement.

"Look here, Bill, this is my lease in Santa Fe Springs."

"Mr. Julian, Mr. Chauncey Julian." A white-gloved bellboy began circulating in the bar area carrying a small metal tray, calling out the name, "Chauncey Julian."

"Here, young man." Julian rose, gave the bellboy a dollar bill, took the card, looked at it and placed it in his pocket. "Dreadful shame. I must leave at once. Some business I'm working on. A radio station deal. Miss Kameal, again, it was a real pleasure to meet you. Bill, don't get up, let's shake to both of our futures. Good-bye for now."

Chauncey Julian proceeded to the lobby at a fast pace looking for the bellboy. He found him standing near the Concierge desk. Tapping him on the shoulder, Julian whispered, "You nit-wit, I told you to page me in the bar in thirty minutes, not ten. You were supposed to begin in the dining room,

< 29 >

and work your way to the lobby, and then the bar."

"I'm sorry, Mr. Julian, I started in the wrong place. It won't happen again."

"Never mind, I like your style. What's your name?"

"Jim Beebe, sir. I work here part time and part time at KHJ as an engineer."

"Listen, Jim, I'll be moving into this hotel in September. I've got some big plans. Stick around. I think I have a spot for you so you can get rid of that monkey suit. Look, I've got to get out of here before my friend and his girlfriend spot me. I'll talk to you later."

"Thanks, Mr. Julian."

"Wasn't that departure a bit abrupt, Bill?"

"Indeed it was, Kathryn. Of course, he never said he was going to pick up the tab. Waiter, our check!"

Bill drove Kathryn back to Hollywood and parked his car next to the flats on Crescent Heights Drive which he had noted the previous Sunday.

"Let's get out and take a look at one of these."

Bill and Kathryn walked into the courtyard of the small complex. Construction was nearly complete.

"I'm going to give them a call on Monday. I like the fact that these units are larger than most apartments and each unit has its own garage. Also, it's close to where you live."

"Do you think that's good for you, Bill?"

"I sure do." Bill held Kathryn's hand. The couple crossed Crescent Heights then turned on Sunset.

"Would you like to come in for a night cap?"

"Sure, Kathryn. This will be our second weekend together."

"Come in, Bill, don't be shy."

CHAPTER 4

"Mr. Feeney, a Mr. Julian to see you."

"Thanks, Maggie, ask him to relax, I'll be right down." Bill Feeney hung up, rose, and descended the stairs to the newspaper's reception area. Chauncey Julian greeted Feeney with a firm handshake. He was smiling and seemed more confident than at Charley 0's Bar several months ago.

He was again dressed impeccably; double-breasted navy blue suit, white shirt with French cuffs, a white silk tie with dark blue polka-dots. Clean shaven, his hair was trimmed shorter than before and was brushed in place perfectly, like a motion picture actor right before a formal scene.

"You are looking prosperous, C.C. What brings you to Hollywood? I haven't heard from you in months."

"I just placed a quarter page ad in your paper for next week. I also ordered some follow-up ads. The same ad breaks next week in all the local dailies. Bill, I'm going to give the folks the opportunity of a lifetime to share in the wealth of Julian Pete."

< 30 >

"Who is Julian Pete, C.C."

"My boy, it's not who Julian Pete is, it's what it is, Julian Petroleum Corporation; chartered in the great state of Delaware. Bill, I've obtained a permit from the California Commissioner of Corporations to issue 600,000 shares of preferred stock and 600,000 shares of common.

This is going to be the opportunity of a lifetime for a lot of the little guys to reap enormous profits. The gushers are coming in, Bill. Out there in Santa Fe Springs, I have my first lease for six acres of the gentle rolling plain, a mound of wealth. I can see them coming in right now."

"Wait a minute, C.C. How are you going to match a six-acre lease hold against Standard Oil Company or Union Oil Company. You know they're fairly competitive. Union Oil is exceptionally strong. Lyman Stewart manages a safe Union Oil Company for the Californians. These are major old time companies. How can you compete with the Signal Hill Boys? What makes you think Julian Petroleum can be a competitive company?"

"Look at it this way, Bill. How many stockholders does Union Oil Company have? I'll tell you, a measly 4,200, mostly old time Los Angeles big shots. I'm going to share the wealth with the little guys. The hoi polloi."

"How certain are you that your gushers are going to come in?"

"You tell 'em 'Wells Fargo.' It's hard to express how confident I am. Listen, Bill boy, I'm having open house at my new offices next Friday, September 29th at 6 p.m., mainly for my staff. My ads hit the papers Friday, the 30th. I want you and your girlfriend, Kathryn—what was her last name again?"

"It used to be Kameal, now it's Feeney. We were married September 2nd."

"Why wasn't I invited?"

"Really no one was invited, C.C. My parents are Catholics. Hers are Episcopalians. To avoid a squabble, we were married in a Catholic Church rectory. My editor was my best man. One of Kathryn's actress friends stood up for her. We're living in my place on Crescent Heights."

"Congratulations, Bill, the same to Kathryn. I have one more stop to make at the Santa Monica Outlook."

"Where is your office, C.C.? Kathryn and I will be at your office warming."

"Seventh and Broadway, Bill, Loew's State Theater Building, Suite 321, 322, 323, and 324. I'll see you Friday." Julian's departure left Bill Feeney curious about the content of the Julian Pete ad. He walked over to the advertising counter.

"Miss Carr, would you read to me the ad copy for Julian Petroleum Company? Did Mr. Julian pay in advance?"

"No, he asked me to bill the Corporation. Let's see. The first line is in one inch caps: 'IT WILL DO YOUR HEART GOOD.' This is the message."

To drive out Telegraph Road to the Santa Fe Springs
field. Prices of leases are jumping fast and oil land in this west-
ward extension of the Santa Fe Springs Oil District is hard
to get. Drive down today and watch the fastest 15 men that

< 31 >

ever stepped into a drilling rig, setting that machinery in place on my first well on my six acre lease there. I'll promise you'll see no lost motion. We'll spud her in sometime this week, and 'Oh Boy', how these 'Birds' will ramble downward. They'll burn the air, that's all.

The big question that you must answer now is, are you going to be with me? Are you going to take a chance on turning a few dollars into a real sized bankroll?

Some wise 'Guy' once said, 'your first thousand is the hardest to get together.' If I don't turn every $100 you shoot with me into at least a thousand, I'll say I know nothing about the oil game, and after all, that's really all I know anything about. Folks, there is very little chance of your going wrong on me, because my greatest ambition in life is to make good on each of these wells. Even if I drill a dry hole, you'll be 'sitting pretty,' because I will drill the second of these two wells on some other sure-fire tract and you'll participate just the same, but I will never call on you for another dime.

Have you stopped to realize I am giving you sixty bills of every hundred that both of these wells ever produced? Not one, but two. The 60% I have divided into 3,500 equal parts and for every $200 you invest with me, you'll get a direct assignment of 1-3,500 part of this production, with one of our leading banks appointed as trustee for you, to collect direct from the pipeline company every 30 days and pay you all each month.

Let's say, 'Here goes nothing,' shoot me the old check for what you have laying around not working and see if I don't bring you home the bacon.'

"Mr. Feeney, he wants his name in one inch caps, 'C.C. JULIAN.' Well, what do you think?"

"It's unbelievable. No, that's not the word, it's outlandish. Let me see the ad for the following week."

"Here, Mr. Feeney. He wants this for October 8th."

Let me see, just 15 days more! And the opportunity to make a nice little wad of dough for a very small investment would have passed out. Folks, this is positively the last offer I will submit to you this year, for I assure you I have my hands full completing the wells I am now drilling and putting in production.

I say, 15 days more for you to take advantage of my offer. Perhaps it will last that long. I'm not so very far from being fully subscribed right now. I know I'm offering you the most legitimate oil investment ever submitted to the public. I'm just sure that you will make several 100% on your money. The odds are 50 to 1 to drill two great wells on my six acre lease which is considered the cream of Santa Fe Springs.

< 32 >

*You have to at least agree with me that there is
big $ in the Oil Game if you get in right, and if I'm not
putting you in right, then there will never be a 'right' ad offer
in this world.*

*I feel that you will give me credit for knowing
considerable for pushing down holes, after hanging up a
'world's record' for speed on my last two wells.*

*You know that I'm drilling on one of the highest
price pieces of oil land in the world.*

*You know that I'm paying 28% royalty to the land
owner, putting up the purchase price myself, and only
retaining 12% of the production.*

*You know that the balance of the production which
is 60 bills out of every 100 produced from my two wells
goes directly to you every 30 days.*

*You know this 60% is divided into 3,500 equal
assignments of production and for every $200 you shoot,
you get a 1-3,500 of 60% from the two wells as long as
they produce.*

*You know that the bank collects from the pipeline
company and pays directly to you.*

*I want you to know 'that if it is the last thing I ever
do, I will successfully complete these two wells.'*

*I also want you to know that I never was sure of
anything in my life more than I am that I will earn you at
least $1,000 for every $100 you throw to me, and I will be
greatly disappointed if it's not twice that much.*

*Now, if you don't know all these things, I can
certainly conclusively prove to you that I'm telling
you the whole truth and if you do not know all these
things — for the 'love of Mike,' why hesitate?*

*Get the word to us quick or you are going to be
out of luck,*

C.C.   *JULIAN.*

"What if the people go for it, Mr. Feeney?"

"It's hard to say where this thing will end up, Miss Carr. I have a confession to make. I hardly know this fellow and I gave him a book about how to start an oil company. This ad does not match my understanding of the book. Make sure you bill him."

Bill Feeney returned to his desk. He picked up the phone receiver, "Maggie, would you ring my wife's dressing room; take a chance—she may be off the set."

Bill Feeney rocked back in his wooden swivel chair, lifting his left foot and resting it in a half open drawer. "Kathryn, I'm glad I caught you. I love you, because I love you. Listen, remember that oil man we met at the Alexandria Hotel?"

"I certainly do, he tricked you into picking up his tab."

< 33 >

"Let bygones be bygones. Listen, he's having an open house next Friday evening. Just his staff, I think. Can you make it. I have an idea, there's going to be a story there."

"Sure, we're not shooting Friday afternoon."

"Great, I'll see you at home about seven."

Bill rested the phone receiver on the hook momentarily, lifting it to his ear a second time.

"Maggie, is Raul Dominguez in?"

"Yes, Mr. Feeney, would you like to speak to him?"

"Yes, please."

"Raul, can you go with me to an oil drilling operation in Santa Fe Springs?"

"Sure, when do you want to leave? It'll probably take well over an hour for us to get there."

"How about right now?"

"Your car?"

"I'll meet you in the parking lot. Bring a camera with a wide angle lens."

"Right." They met in the parking lot.

"Bill, let's grab a bite at Olvera Street."

"Great idea!" Bill drove east on Hollywood Boulevard to the junction with Sunset. At Glendale Boulevard, he remarked: "Looks like Sister Aimee's Angelus Temple is about completed. She rode in a special Pullman on the train when I arrived in town in February. I read in the paper the other day that she intends to have dedication ceremonies New Years Day."

"Bill, I was in that group of photographers at the train station. Remember seeing me?"

"No, I was mesmerized by the whole scene. It was my first unofficial meeting with a celebrity other than old Death Valley Scotty."

"Where did you meet him?"

"In Barstow. He was on his way to meet his partner and his partner's wife at the Alexandria Hotel."

"Bill, make a right turn and park on Spring Street. We can walk over to Olvera Street." They sat in an outdoor but shaded area at the "Cielito Lindo." A trio was singing Norteno music at a restaurant nearby. "Raul will you order for me?"

"Sure."

"Camarero, dos platos de Pueblo combinaciones."

"Si, Senor."

After lunch, they walked toward the Plaza.

"Bill, if prohibition is abolished, this old building could reopen as the first winery in Los Angeles. It's the Old Pelanconi House—built in the Fifties—the first brick building in the city." The two men crossed Main Street in front of La Placita, the Plaza Church. "Let's make a visit, Bill, it will do us good." Bill followed Raul into the dark interior of the Plaza Church, an out mission built for the local Indians by the Franciscan Fathers, who at the time of the construction of the church, were headquartered at the Mission San

Gabriel, 20 miles to the east. Bill and Raul dipped their hands into the Holy Water font, blessed themselves, genuflected, then knelt in silent prayer. To the left of the entry was a large crucifix illuminated by candles held upright in a bed of sand. The main altar, wooden, and gold encrusted stood in front of a floor to ceiling ornate wooden reredos designed to display paintings of several Spanish saints. On top of the reredos, in the middle at the highest place of honor, was a painting of Our Lady of Los Angeles de Porciuncula. Mary, the Mother of Jesus wearing a crown hovered in the clouds, holding the infant Jesus supported by angels.

After their brief prayer, Raul and Bill rose, genuflected a second time, once more blessed themselves with Holy Water, and walked through the Church courtyard toward Spring Street.

"What did you pray for, Amigo?"

"I prayed that my wife, will get pregnant and that we will have a happy family life."

"After seeing your wife, I am sure you will be able to help the Lord take care of that problem."

"O.K. Raul, that's enough. What did you pray for?"

"I prayed that one day California will have a governor with a name like mine, Gonzalez, Duran, maybe even another, Pico. Pio Pico, our last Mexican Governor built that hotel across the street from the Plaza Church. By the way, Bill, did you notice the well-dressed man in a dark suit kneeling in the pew in front of us?"

"Yes, I thought he looked very distinguished."

"That was Joseph Scott, a prominent attorney. He is very close to the bankers in this town. The Italians trust him because he is Catholic. The next time I see him, I'll introduce you. He is a close friend of my father."

"Thanks, Raul."

"Make a U-turn and head East on Macy Street. We'll take a short cut on the new Macy Street Viaduct. When you get to Soto Street, turn right and then left on Whittier Boulevard."

They arrived at Whittier Boulevard and Santa Fe Springs Road at 2:00 pm. and headed west towards the Santa Fe Springs Oil Fields. The odor of raw oil from the refineries was pungent.

"When we get to Telegraph Road, turn right?"

"Raul, look at those Indian head signs."

"There's a derrick being built. Let's stop and see what's going on. Don't bring your camera yet. Howdy, partner, my name is Bill Feeney from the Citizen News."

"Roy Evans, I'm in charge of field operations for the Santa Fe Chief Oil Company. We have a ten acre lease here. This is our first derrick. Over yonder is Union Oil's Bell Number One, on the Bell Ranch. There on the rise is J. Paul Getty's Nordstrum Number One."

"Can you tell me where the Julian Petroleum lease is located?"

"Sure, keep going west on Telegraph Road to Pioneer, make a left turn, go south to Little Lake Road, make a left and you'll see the derricks on the north side of Little Lake Road."

< 35 >

"Thanks, Roy. If my cameraman takes a picture of your crew, is it all right?"

"Sure."

"Raul, will you please get your camera and take a close shot of the crew building this derrick, and get their names" Both men returned to Bill's car. Raul returned to the construction site with his equipment. Bill sat in the driver's seat taking notes. "O.K., Raul, let's go over and take a look at Julian Petroleum."

As they turned south on Pioneer, Bill observed, "I guess we could have hiked from the Santa Fe Chief to Julian's lease. There are two derricks on the north side of Little Lake Road. I'm going to turn left on Fulton Wells Avenue. Yes, indeed, the Julian Petroleum Corporation. Looks like he has two wells producing and two more being drilled. Raul, get some shots of the men at work and get their names. And take a wide angle shot showing the Julian Oil sign on the building."

Suddenly a hissing noise to the east startled them. A crew of men working on a drill rig were running from the rig in several directions. The hissing noise increased to a roar. The drill rig quivered. Then, it blew. Earth, drill rig parts, timbers, pipes, and dust flew into the air. A cloud of dust drifted with the wind northward toward the Santa Fe Chief leasehold. The hissing sound continued. The gases blew upward, creating a pockmark crater on the gently sloping land. Bill shouted, "Are you all right, Raul?"

"I'm O.K. Let's get out of here! This dust will ruin my cameras. That thing sounded like an exhaust pipe from the nether world."

CHAPTER 5

"Is Bill Feeney in? This is C.C. Julian calling. Is that you, Kathryn?"

"Yes, C.C., I'll call Bill and thanks again for the ten units of Julian Petroleum stock. Hopefully, it will help Bill Jr. with his college education."

"Think nothing of it. Tell me why nights are lonesome, tell me why days are blue, tell my why all the sunshine comes just at one time, when I'm with you. Kathryn, I'm having a ribbon cutting ceremony at my new gasoline station in Cucamonga on the 4th of July. I would like you, Bill, and your baby to be my guests. It's a Wednesday, will Bill be working?"

"I don't know, I'll let you talk to him. Goodbye C.C., and thanks again for the gift."

"Bye, my dear."

"Hello C.C., what's up?"

"I'm having a ribbon cutting ceremony at my new gasoline station in Cucamonga. How about you, the Mrs., and your son riding out with me. I just bought a new Pierce Arrow."

"I'm off that day and we'll make it, but I'd rather meet you at your office and follow you out in my own car."

"Bill, don't meet me at the office, I've taken one of the bungalows at the Ambassador Hotel. Go to the lobby and have them ring me. They will direct you to my bungalow."

< 36 >

"C.C., should we bring a picnic basket?"

"No, we're going to have free Julian burgers and orange juice at my new diner near the gasoline station."

"OK C.C., we'll see you Wednesday morning at nine am. "

Chauncey Julian sat at his desk in the executive suite of his third floor offices overlooking the corner of Seventh and Broadway in downtown Los Angeles. He checked his calendar and tore off June, 1923. July promised to be a banner month at Julian Petroleum; the tenth of the Julian gasoline stations would be opening on the San Bernardino Avenue in the Rancho Cucamonga Wine country.

Jim Beebe entered the office. "Would you like some coffee, Mr. Julian?"

"I think so, let's go down and have a cup!"

Chauncey Julian and Jim Beebe took the elevator to the second floor. A few steps to their left, they entered the brightly lighted Julian Snack Shop.

"Good morning, Mr. Julian." Ray Wong, the manager greeted him. "Good morning, Jim."

"Good morning to you Ray, how is business?"

"Very good, Jim—very good—ever since Mr. Julian comes here—very good. Coffee, Mr. Julian? You, Jim?"

"Yes, coffee for both of us."

Julian placed his right arm on Wong's left shoulder. "Is business slow, Ray, since our stock sales campaign is not quite as active as it was?"

"Slowed down Mr. Julian—little bit—people still come-see you—we still make good business."

"Good Ray. We only do good business."

Julian and Jim Beebe took their mugs of coffee to a table. "Jim, have you completed arrangements to rent the Glenn Ranch in Lytle Canyon for our party after the ribbon cutting?"

"Yes sir, Mr. Julian. We booked in the ranch exclusively."

"Did you order wine from the Cucamonga Winery?"

"I tried to, sir, but no one answered the phone so I sent a purchase order. Hopefully, they will have our order ready July 4th."

"Jim, you better hire a photographer for this ribbon cutting. I question whether any newspaper will bother sending staff to Cucamonga, especially since I've been cutting down on advertising."

"Yes sir."

"Did you arrange for the Sedgwick sisters to hold each end of the ribbon when I cut it?"

"Yes sir. They will wear their cowgirl outfits and will fire their .38's after the ribbon is cut."

"Did you line up the Julian band?"

"Yes sir, they will be at the station, but not at the ranch."

"Could you find an official to represent the community?"

"Yes, Mr. Julian, the Cucamonga Constable promised that he would be there. He'll wear his uniform and lead us in his police car up to the Glenn Ranch."

"Is the Julian Diner ready to go?"

< 37 >

"Yes sir, we have a chef and two waitresses. We'll be serving free hamburgers and orange juice. The Euclid Orange Association has given us a special deal on oranges and lemons. All we have to do is put their signs inside and outside the diner in prominent positions."

"Good! Who is on the celebrity invitation list besides the Sedgwick sisters and they are, for sure, right?"

"Yes sir. Let' s see, besides some of the larger stockholders, we've invited Norma Talmadge, Louella Parsons, John Robertson, Howard Hughes, Charles Chaplin, Wallace Beery, and Hoot Gibson."

"Why do we want Hoot Gibson? Isn't he the cowboy who is promoting Arrowlube Oil Company?"

"Yes sir, but for $100, he will ride his horse into the main ranch area during the barbecue and put on his rope throwing act. He plans to lasso the Sedgwick sisters. He's bringing along a singing cowboy named Fred Gilman. I don't think anybody will pay attention to his past work for Arrowlube Oil Company. Besides, I know he hasn't done anything for them lately."

"Jim, how many people will be spending the night?"

"I don't really know, Mr. Julian. That depends on how well the party goes and how many hangers-on the celebrities bring. There's usually a pretty big crowd when the food and drinks are free. I know our office staff, their wives, and girlfriends will stay the night, and I assume you and your wife will stay, as well."

"Yes. But remember we must be back at the office by ten or eleven." They finished their coffee and Julian saluted the coffee shop manager who gave his usual parting remark, "Thank you, Mr. Julian...Mr. Julian no pay nothing—good man."

Chauncey Julian and Jim Beebe returned to the office. It was quiet, a Saturday. Julian sat near his small roll top desk, picked up a pad of lined paper, and began writing the speech for the ceremony.

"Say Jim, send an invitation for the ribbon-cutting and the party to that oil column writer for the L.A. Times. He never puts my name in his column and maybe we can shake him loose for a few lines."

As the sun rays filtered through on July 4th, 1923 in Cucamonga, California, the early morning fog drawn inland by the prevailing winds through the Santa Ana Canyon was dissolving. Chauncey Julian and his wife arrived at noon in his gleaming Pierce Arrow. Bill Feeney, in his Auburn, his wife, Kathryn, holding their baby in a white wicker bassinet, followed Julian into the parking lot, crossing the boulevard near the stop sign at the corner of Archibald. Groves of lemon and orange trees dominated the sloping terrain. The gasoline station parking lot was crowded with Julian's employ-ees, corporate officers, and well wishers. The two women in cowgirl outfits stood near a large red ribbon stretched across the inner parking area near two gas pumps.

Chauncey Julian parked his car in front of the grand opening ribbon so that his car would be the first one to enter the station for service. Bill Feeney

< 38 >

parked at the curb on the boulevard near three recently constructed homes. The Cucamonga Constable in a tan uniform held a pair of wooden scissors. A photographer, a late arrival, was hastily setting up his tripod and camera. Julian strode to the center of the ribbon. Two youngsters, local children, properly recompensed by Jim Beebe stood on either side. The crowd broke into applause. Chauncey Julian raised his right arm aloft for silence and began his dedicatory speech. Bill Feeney led Kathryn, carrying the baby in his bassinet, into the gasoline station.

Kathryn whispered, "Bill, why didn't Julian ask his wife to stand there with him?"

Bill, cupping his hand to his mouth, responded in a whisper, "He told me that he does not want his wife to be involved in his business affairs."

"Ladies and gentlemen, you have in the recent past witnessed the fact that our gushers in Santa Fe Springs have made money for all of us. I could tell you plenty about the holdings of Julian Petroleum Corporation today, and probably enough to jar the most conservative investor in California loose for a piece of change. But I guess I told you plenty from day to day for the past months in my newspaper advertising. I'll just add "a word to the wise is sufficient." By December of next year, we are determined to open more than 30 Julian gasoline stations. We now have more than 40,000 people who believe in me. Julian Petroleum Corporation stock has 200,000 shares of preferred outstanding with a par value of $50 and an equal number of common shares outstanding. I know I'm right! I know they can't stop me if you will just believe in me and if you're strong enough to just take a chance. I'm on a winning streak and I'll make you dizzy with profits. You know folks, I didn't fall off a Christmas tree myself. I'm trying my best to gain your confidence and hold it. And the only way I can do this is by making you money. If I lost you a dollar, I would expect to be as welcome as a skunk at an afternoon tea. I believe I know this oil business from soup to nuts and I 've given all of you a chance to share in our profits by my stock offers. I promise to play square and I stake my reputation as an oil man to successfully finish what I have started. And now, the ribbon cutting!"

Julian moved to the center of the ribbon. He took the large wooden scissors from the Cucamonga Constable and simulated a cut as the Cucamonga Constable slit the ribbon with real scissors. As the strands fell to the sidewalk, the Sedgwick sisters, Josie and Babe, each fired three rounds from their six shooters. The crowd burst into applause as the Julian Band played "Over There."

Jim Beebe raised both hands and shouted, "Now, everybody, to the Julian Diner for free Julian burgers and orange juice!"

As the crowd dispersed to the diner, Kathryn Feeney remained inside the station office feeding her baby and gently rocking the bassinet. A dark haired woman entered the office and approached Kathryn.

"I am Mary Olive, Chauncey Julian's wife. Your baby is darling! What's her name?"

"It's a boy, William Warren Feeney, Jr. My husband is an acquaintance of Mr. Julian. I'm Kathryn. I told my husband we can only stay for a short time as I want to be home before sundown. I know how these celebrity

< 39 >

parties go on till the wee hours. By the way, how did you meet your husband?"

"I met him in Winnipeg, Canada in 1910. He purchased a clothing store with my father's help. My father bought his clothes there previously. Our first daughter, Lois, was born in 1914, then Frances a year later. In 1917 he closed the store and came to the United States. He first went to a place called Bakersfield where he worked in the oil fields. He got tired of that kind of work and returned to Canada. Then he got restless and came back to the United States, first to Los Angeles, then to Texas. We corresponded regularly and told me he was very lonely. He wrote about his enthusiasm for selling oil stock and for the last two years we have corresponded weekly. He asked us to come down last month and arranged for us to live in a bungalow at the Ambassador Hotel. He buys me expensive clothes and jewelry, treats the children wonderfully when he sees them, but I know little about his business. He tells me I should enjoy California and leave business matters to him. What line of work is your husband in?"

"He's a newspaper reporter and I'm a former actress. As long as my husband can support us, I'll probably not work. Why don't you ride out to the Glenn Ranch in our car. Have you made many friends since you arrived in Los Angeles?"

"Not really. Chauncey is so busy at his office that he returns to the hotel late in the evening. On weekends we go to parties but I don't really know the people that well - mainly show business people. Also Chauncey is trying to buy a radio station and if he is able to complete the deal, he wants me to manage the office for him. I hope I can meet some friends that way."

"Bill, come in here, this is Mary Olive, C.C.'s wife."

"Nice to meet you. C.C. never told me that you were in town."

"He's so forgetful. You have a very sweet child, Mr. Feeney."

"Please call me Bill."

"Bill, I've asked Mary Olive to ride to Glenn Ranch with us. Is that all right with you?"

"That's great. C.C. asked me to follow Jim Beebe over to the Cucamonga Winery to pick up cases of wine for the party."

"Bill, I'll tell Chauncey that I'll be driving up with you. He asked if I would accompany the Cucamonga Constable. The Constable is to lead the cars up the canyon. I'm sure C.C. will allow me to be with you two and the baby."

"Bill, darling, I'm not going up that canyon and tangle into a Hollywood orgy for very long. We'll take up the load of wine, pay our respects, and head back home."

"That's what I had in mind Kathryn, don't worry."

Kathryn got in the back seat with the baby and closed the door. Mary Olive sat up front. Bill started the Auburn and followed Jim Beebe in his Dodge Brothers sedan. The two cars, after a boulevard stop, made a right turn on Archibald and then to a railroad crossing, making a left turn on a dirt road past the Rancho Cucamonga Santa Fe Railway Station.

"Kathryn, when I arrived on the Santa Fe Train February of last

< 40 >

year, our train made a special stop at that station. A fellow named John Finlay got off to place an order at the winery down the street."

As the cars approached the winery, Beebe slowed down at the west entrance which was blocked by a boxcar on a side rail adjacent to a loading dock. He continued past the northern wall of the winery, eastward. Brightly painted signs on the upper portion of the concrete wall displayed wine bottles and grapes. In the middle of the building was a sign:

Bonded Winery #1
Padre Vineyard Co.
Vau Brothers since 1870

The cars entered a driveway near loading docks and vats where wine grapes were unloaded. Beebe parked his car in front of the main entrance to the winery next to a large truck with a "BONDED WINERY #1" sign on the cab door. Bill parked next to Jim's sedan. A few yards to the south under an arbor, protected from the bright sunlight by lush green flowering vines, was a group of about fifty people. Steaks and chickens were being barbecued on a grill about the size of a water well made of concrete and river bottom rocks. A mariachi group played in the background.

"Bill, let's get the wine loaded. If it won't all fit, we can put a case or two in the trunk of your car."

The two men entered the winery. The murky desolate interior echoed their footsteps. Empty wine barrels were strewn about. A rat retreated under a vacant counter. The odor of decay and rancid fruit permeated the air. Dust particles floated in the stream of sunlight from the open door like fine grains of fool's gold. An old Mexican stood in the doorway. "Que desea usted?"

"I'm Jim Beebe from Julian Petroleum Corporation. I mailed the winery a purchase order last week."

"Esta cerrado, close it."

"Who closed it?"

"La policia."

"Just a minute, Jim. Senor, a donde va el viejo Aleman? The old German, El propietario. The owner. Where is he?"

"El viejo muerto. Los bootleggers. Esta perdido en el desierto."

"Jim, we aren't going to get any wine here. I think he is trying to tell us the old German is lost in the Sheriff's desert graveyard, whereabouts unknown. Let's go."

They returned to their automobiles and joined the caravan. The timing was perfect. The cars led by the Cucamonga Constable were heading north on Archibald toward the base of Cucamonga Peak and, joined by Bill Feeney and Jim Beebe, proceeded north through the lemon orchards, turning east at a road sign indicating Lytle Canyon eight miles. The roadway was shaded by eucalyptus trees which served as windbreaks to protect the lemon orchards. The roadway abruptly ended at a street called Haven facing seemingly endless grape orchards. Turning left on Haven the caravan arrived at Highland, and continued eastward along a road which served as the northern boundary between the cultivated vineyards to the south, and chaparral, sloping upward to the base of Cucamonga peak and the San Savaine Ridge. At Lytle Creek Road, the Cucamonga Constable motioned his followers to make a left turn beginning the ascent into Lytle Canyon. The road narrowed as it crossed a dry gulch. The canyon was framed by steeply rising pine covered mountains to the west, with smaller hills to the east, the lower portion of the Cajon Pass. The road hugged the edge of a river bed with flowing water which had been totally dry several miles down stream. The caravan passed a village of small cabins and halted at a closed thick wooden gate set on steel rollers. A guard shack stood inside the securely fenced area on the east side of Lytle Creek Road. On the opposite side of the road, the creek flowed through a grove of sycamore trees. At a word in Spanish from the constable, a Mexican unlocked the gate from within. It easily slid northward on its steel track opening to a width which could accommodate two automobiles. The constable led his followers into a parking area in front of a Spanish style two story rancho. Several limousines were already present as the Julian ribbon cutting attendees entered the parking area. The constable got out of his vehicle and strode to a shaded area under a portico which covered the northern and western portion of the rancho, a shelter from the afternoon sun. Chauncey Julian motioned his well wishers to join him at the resort's main entrance. The wooden door to the main salon of the rancho contained a hand carved inscription: "Welcome Guests and Friends."

As the final stragglers joined the crowd, the constable announced, "Welcome to the Glenn Ranch! The staff will show you to your cabanas. The swimming pool entrance is either through the bar, off the main salon or at the east end of the ranch house. Tennis courts are available as well as horseback riding. Mr. Julian asked me to invite all of you to a barbecue dinner at 6:00 p.m. It will be held in the grass area just east of the swimming pool. Thanks to the generosity of Mr. Julian, the entire Glenn Ranch is yours to enjoy. Remember, checkout time tomorrow morning is before 11:00 am. You folks won't need me to show you the way back. If you try to go up the canyon you'll be following a trail used by pioneers and not in very good shape for automobiles. Have fun."

"Kathryn, perhaps you should take the baby inside. It's too hot here."

"Bill, I'm going to find some milk for the baby and ice for the trip back to Los Angeles. I'll be near the bar unless I find a shady spot by the pool. If I recognize anyone, I might go out."

Chauncey Julian took his wife to one of the red tile roofed cabanas with a sign above the door "Trail Finder."

"Mary Olive, I'll have your suitcase brought over here as soon as things settle down. There are cool soft drinks in the icebox. If you feel like sitting near the pool, walk on over. I'm going to circulate. I'll come back for you before the barbecue."

He walked over to a mirror and combed his hair. "Mary Olive, I'll be up late tonight and will be staying over at the main ranch house. I'm going there now to shower and put on a fresh shirt. I want to look my best. My people respect me for the way I take care of myself. Respect leads to trust.

< 42 >

I need that."

"Chauncey, I love you, you know that don't you?"

"Yes, Mary Olive."

Julian kissed her on the forehead. He then walked across the grassy area toward the main house. Cowboys were unloading a horse from a trailer. One with a large white ten gallon hat, bowlegged, wearing leather chaps, barked orders to two others who were unloading tack from the front part of the trailer. That must be Hoot Gibson, thought Julian. One hundred bucks.

Kathryn asked the bartender for some cold milk and then carried the baby to the entry way leading from the bar to the pool side. A number of Julian's guests were already pool side. Several were in swimsuits, others wore robes despite the heat. Kathryn recognized no celebrities other than Mary Miles Minter, the star of the recently released, *The Trail of the Lonesome Pine*, who more or less disappeared from the Hollywood scene after the murder of her former director, Desmond Taylor. Behind Kathryn, in the main salon, a rustling and clatter could be heard as a few groups of men began card games.

"Sir!" Kathryn turned to the bartender.

"Call me Oscar, mam."

"OK, Oscar, do many Hollywood folks come up here on weekends?"

"Oh yes, Miss, a lot of them, especially the child stars and their parents. They like the ranch atmosphere. Last weekend Baby Peggy, her sister, Louise, and the parents spent the weekend with us as did Adolph Menjou and the tennis star, William Tilden. But on this holiday Mr. Julian has rented the entire facility and, except for a few wealthy people who are staying more or less in seclusion at the Glenn Family Home, Mr. Julian's guests are the only folks here."

"Oscar, would you please put some ice cubes in some sort of container so I can keep the baby's milk bottles cold during our trip back to Los Angeles? My husband and I are not going to stay for the barbecue."

"Sure mam. I'll use an empty ice cream carton and have it ready when you are."

"What's up darling?" Bill Feeney entered the bar sleeves rolled up and perspiring. "Beebe and I carried two beer kegs up to the barbecue area. They have some tubs filled with ice. When do you want to scram?"

"Any time. The baby is sleeping. Bill, meet Oscar the bartender."

"Hello, Oscar."

"Care for anything from the bar?"

"What do you have?"

"Soft drinks only."

"Give me anything cold. Any celebrities out there, Kathryn?"

"Just one of any note and she is dropping out of sight. They say she's getting involved with investments. She must have bought some Julian stock. Where is Mary Olive?"

"C.C. reserved a cabin for her above the grassy area. Hoot Gibson has his horse tethered up there with the riding horses owned by the ranch. Some of Julian's guests asked to ride and the ranch hands brought out some

< 43 >

horses from the stables."

"Did you tell Julian we were leaving early?"

"Yes, and I thanked him. Let's go!"

"Bill, get the container with ice from Oscar and give him a good tip."

Meanwhile at the barbecue area, Fred Gilman, the singing cowboy with three musicians entertained the crowd. Soon a clang of a metal triangle struck by the head chef brought the partying Julian fans to the barbecue of beef, pork, chicken, beans, and salad. Chauncey Julian presided regally on the porch of a cabana with Mary Olive at his side. Horseback riders returned from the upper canyon and tethered their horses on a pine tree stripped of its bark, supported parallel on each end by four foot posts.

The singing cowboy and his musicians moved to a small stage on the porch of one of the cabanas near the horses. As Hoot Gibson mounted his rodeo show horse, the musicians played "The last Longhorn." Fred Gilman gave a cowboy holler and Gibson's horse reared into the air, pawed as if swatting flies then raced in a circle around the grassy area in front of the crowd. The Sedgwick sisters, in cowgirl outfits ran to the center of the grassy area. Hoot Gibson then circled them at high speed twirling his lasso. He reigned in his horse abruptly and hurled his lasso. It circled in the air and gently caught the smiling Sedgwick sisters, cinching them around their waists. Once again, Gibson's horse reared in the air, twirled in a circle and returned to all fours. The crowd cheered and applauded. Like a true showman, Chauncey Julian strode to the center of the grassy area, shook hands with Hoot Gibson, and removed the lasso from the Sedgwick sisters, while Fred Gilman sang, "Lily of the West."

Around midnight, Julian sat alone at the ranch house bar. His friends and followers had retired as had Mary Olive. Hoot Gibson, the Sedgwick sisters, Fred Gilman and his musicians left the ranch and the canyon to return to Hollywood. Sipping from his flask of bourbon, a tired Chauncey Julian was alone with his thoughts except for the bartender.

"How are you feeling, Mr. Julian?"

"Not too well, Oscar. Julian Petroleum has some problems. The price of gasoline is falling and my respectable competitors don't want to acknowledge that I exist. Last month a postal inspector named Madeira wrote a questioning letter to one of my stockholders. I filed a damage suite of $100,000 against him and the government. They had the gall to send out a special Assistant Attorney, General David Cahill, from Washington to reply to my suit. I won't stand still for this harassment. Cahill told newspaper reporters that Postal Inspector Madeira simply sent a letter to a private citizen and blames me for making the matter public. He claims my suit lacks any basis. But they're not going to stop me. I feel as though my enemies are out there trying to take over my business."

Polishing a glass with a dishcloth, Oscar replied, "Mr. Julian I'm going to tell you a little story.

Old Silas Glenn developed this ranch, built the buildings, planted the apple orchards, brought in the livestock, but Old Silas passed away in 1878. His widow tried to fill in his shoes, but it was too much for her. She

< 44 >

turned over the management of the ranch to her daughter and her daughter's husband, Mr.and Mrs. James Applewhite from Mississippi. Two of the widow Glenn's sons John and Silas were married and lived in Bloomington. They often told neighbors that someday they wanted to take over the Glenn Ranch. They resented their mother turning over the ranch management to James Applewhite. On June 23, 1893 John and Silas Glenn Jr. came up here with the intention of having it out with James Applewhite and his son, Oliver. The Glenn boys couldn't find Oliver so they hid out near the lower ranch gate waiting all night for him to come.Oliver had been forewarned that trouble was brewing and entered the ranch over the Cajon Ridge rather than coming up through Lytle Canyon. Around ten a.m. the next day June 24th, the Glenn boys who were armed, returned to the ranch house and asked for Oliver. They said that they knew he was somewhere about. On being told that he was not there they started a quarrel with James Applewhite. While their own mother and sister did their best to calm them down, both of the Glenn boys threatened Applewhite as to what they would do.

Oliver, armed with a revolver came up from the barn, where he had been hiding, to stand by his father. John Glenn catching sight of him exclaimed, "Take that, damn you," and shot as he spoke. Those were the last words he spoke, cause the sixteen year old boy was a split second first and his uncle died with a bullet in his heart. In a few seconds, the father ducked into the house, picked up a shotgun, came through the front door, and shot Silas Glenn, Jr. with a charge of buckshot. Silas died on June 26th. Both the Applewhites, James and his son, Oliver, were exonerated at a coroner's inquest held here on the ranch on June 27, 1893. The point of this is Mr. Julian, don't let people take what's yours." Chauncey Julian rose, placed one arm on the bartender's shoulder, shook his hand saying, "Thanks partner."

Thursday morning Chauncey Julian arrived at his office at the Loew's State Building at 11:30 am. He was greeted by Jim Beebe and other members of the Julian Petroleum staff.

"Mr Julian, may I see you alone for a minute?" "Sure Jim, come on in my office."

"Mr Julian, at 9:30 we received a telephone call from Edwin Daugherty, the California Corporation Commissioner. He told me that next Monday at 9:00 a.m. sharp he and his staff members are coming to our office to examine our books."

Julian said nothing. He turned in his swivel chair and stared out onto Seventh Street. Jim Beebe stood motionless. Then Chauncey Julian turned slowly facing him. He then got up and went to the hat rack, removed his cane and hat, and turned to Beebe, "Jim pack all our books in cardboard boxes. Use the freight elevator and take them to the loading dock in the alley by 2:00 pm. Meet me there and we'll load them in our cars. Pay the staff a month's salary in advance and tell them to keep their mouths shut. I'm going over to the hotel and pick up Mary Olive and the children. We're going to Las Vegas, Nevada. Tell the staff we'll be in touch with them when we get there. We'll be running the company from temporary headquarters. And tell

< 45 >

the staff not to say anything to that fat Irishman. Let him figure out what we're doing. I'll see you at 2:00."

"Good morning, Sir. My name is Madeira. I'm a United States Postal Inspector." Madeira displayed his badge and credentials.

"Yes Sir. What can I do for you?"

"Yesterday I watched you take photographs of Chauncey Julian at the opening of his gasoline station in Cucamonga. You were so employed, were you not?"

"Yes, Sir."

"Have you prepared the prints yet?"

"No, Sir. They were in no rush."

"How many photos did you take?"

"Four or five—a couple during his speech then the ribbon cutting."

"Good! Please prepare extra copies for me. Here is a franked envelope. Mail me a set. My address in Burbank is on my business card. Enclose a bill to the U. S. Postal Inspection Service."

"Yes, Sir. They'll be in the mail tomorrow."

CHAPTER 6

"Good evening, gentlemen!" Mr. Packard, the manager of the Alexandria Hotel, was summoned by the registration clerk who failed to locate reservations for two newly arrived businessmen.

"Mr. Packard, I am sorry to disturb you, sir, but Mr. Lewis and Mr. Bennett tell me they have reservations for two suites and I have no record, sir."

"Mr. Lewis and Mr. Bennett, we are delighted to have you stay with us. I am certain something can be worked out."

"I hope so, sir, Senator William H. King called me while Mr. Bennett and I were working on establishing oil terminals and distribution facilities in the state of Washington and Western Canada. I told Senator King we would be in Los Angeles in early August to investigate a deal with the Edward L. Doheny Company. Senator King told me to be sure to stay at the Alexandria Hotel as the owner is his close friend. Your hotel appears to be magnificent. Mr. Bennett here is my close personal associate. We trust that you can accommodate us, Mr. Packard."

"Of course, Mr. Lewis. There must have been some mixup. Front! Is your luggage with you, Mr. Lewis?"

"No, Mr. Bennett and I rushed down from Portland with only these two bags. Our luggage will be following us in a few days."

"Certainly—of course. Please both of you sign the register. Thank you very much! Please give Senator King our best regards. When he is in town he always stays with us."

"To be sure, Mr. Packard. Mr. Bennett and I will find our own rooms. The keys, please?"

"But, of course."

< 46 >

S. C. Lewis, balding, heavy jowled with dark hair, forming a crescent around his deeply receding hairline, picked up his leather suitcase, and was followed to the elevator bank by his neatly dressed business associate, Jack Bennett. Jack Bennett, a younger man with heavy eyebrows and piercing eyes with brown bags underneath, left the hotel manager with the assumption that Jack Bennett was the trusted business lieutenant of the wealthy and influential oil man, S.C. Lewis.

"Jake, tidy up and come over to my room. I want to place a phone call to George Johnson and find out when we should test the ground at Julian Petroleum."

"Sure S.C., give me ten minutes."

Lewis entered his hotel suite, placed his suitcase on a luggage rack in the closet, opened it, took out a sweatshirt, a pair of corduroy trousers, and a pair of old slippers. He changed his clothes, placing his dark blue pinstriped suit, silk shirt, and tie on his bed.He moved to a dressing table, picked up the phone receiver and called for maid service.

"This is S.C. Lewis in 728. Please pick up a suit and shirt from 728 and also please pick up a suit and shirt from my business associate in 729. We need them cleaned and pressed by 9:00 a.m. This is an emergency! Our luggage was misplaced during our trip here from Oregon. Thank you, very much, my dear." He then asked the operator for room 729.

"Jake, get your suit and shirt ready for maid service right away. We need to look our best tomorrow morning."

Lewis unbolted the door between the suites. Jack Bennett did likewise and prepared his clothing. A team player, Jack Bennett was prepared to move rapidly and efficiently when S.C. Lewis started some action. A maid knocked at the door of room 729. Jack Bennett had also changed into casual attire. Lewis opened the door and handed the clothing to the maid. She wrote a receipt.

"Here you are, sir. We will have your suits and shirts ready for you by 9:00 a.m. tomorrow. There will be an extra charge for the quick service." "That's fine, my dear. Just take care of it!" "Yes sir."

"Sit down Jake, I'll see if I can get hold of George Johnson."

S.C. Lewis picked up the telephone, opened a notebook, and gave the telephone number of a Los Angeles stockbroker, George Johnson, to the hotel operator.

"George Johnson, please."

"This is George Johnson."

"This is S.C. Lewis. George, Jake Burman, and I just got into town. We're staying at the Alexandria. By the way, Jake is going to be using the name Jack Bennett here because he had a little problem in New Jersey which he settled but still wants to be low key. What's the status of Julian Petroleum? I read in the newspaper that Chauncey Julian moved his operation to Las Vegas."

"S.C., good to hear from you. You certainly got here quickly! I'm glad you called. I think it's about time we started making some moves. I think our timing is perfect. Julian got himself into trouble with E.M. Dougherty, the California Corporation Commissioner. Dougherty has accused Julian of

< 47 >

violating the Blue Sky Law of the Corporate Securities Act. It's kind of old stuff. But Dougherty claims Julian started his newspaper ad campaign to sell stock and started buying property before he had the requirsd permit to sell the stock. Julian applied for a permit. Dougherty wanted to look at the books. But Julian packed up and headed for Las Vegas along with his board of directors. I got word from one of the board members that Julian instructed the board to sell him 100,000 shares which they did. That deal was legitimate but Julian got bored in Las Vegas, came back to Los Angeles, and started to sell some of those shares. Dougherty got word that Julian was back in town selling his own personal shares. Dougherty then raided Julian's office, seized his books, and had Julian arrested. Julian then promised to obey the California Securities Law so Dougherty backed off, returned the company books, and issued Julian a permit to sell Julian Petroleum stock once again. That's all Julian needed. He started reloading, offering owners of unprofit able units of his later syndicates the right to exchange each unit for two shares of Julian preferred providing they would buy for cash an equal number of Julian shares."

"George, I think Julian is desperate. Tomorrow I'd like to borrow your car. I want to take Jake out to Santa Fe Springs to take a look at the Julian Petroleum facilities. Call Julian and tell him some wealthy folks are in town who are spreading the word that they have cash and are interested in Julian Petroleum - the whole works!"

"S.C., when I heard you were leaving Portland, I called Julian's office and spoke to Jim Beebe his office manager. Julian is in New York trying to sell some of his stock and won't be back until Monday."

"Good. That will give me some time to get acquainted with the operation without him wondering what I'm doing. Call his assistant and arrange for a meeting in the dining room at the Biltmore Hotel. Make it for noon. If there are any hitches, call me."

"Very well, S.C., you and Jake, I mean Jack can pick up my car tomorrow at the Pacific Coast Stock Exchange. Tell Jack to meet me on the floor and I'll give him the keys."

"Thanks George, I'll be hearing from you—wait a minute Jack wants to ask you something."

"Hi George, nice talking to you again. When we start working on this Julian deal, we may get into some difficulties. Are there any politicians or judges and the like around this town who are vulnerable?"

"Jack, stay away from the Feds. Also, the State people are not very cooperative. If I need them, there are some local politicians who probably would be cooperative. I don't like to talk about such things on the phone. I'll talk to you about it some other time."

"OK George, so long for now."

Jake turned to S.C.

"Well, Jake, are you ready for the ride?"

"S.C., this town had too much money floating around. It needs a cleansing. For a starter, we need some new threads. Have your attorney friend, Senator King, arrange for a line of credit at some bank, so we don't have to ring for room service to get our suits out of a pawn shop."

< 48 >

"First thing in the morning, Jake."

"Hi Jim, is C.C. in? I'd like an interview for a piece I'm doing about C.C.'s trip to New York."

"Sure, Bill, I'll buzz him. How are Kathryn and your son?"

"Fine. We bought a place in Lankershim, near Toluca Lake. We needed some room for Bill to play and Kathryn is doing a few things at First National Studios. It's walking distance."

Jim Beebe picked up the intercom phone and called Chauncey Julian in his private office.

"Mr. Julian, Bill Feeney would like an interview." "Sure, have him come in."

Bill Feeney entered Julian's office. He noticed that the old glow was missing. Julian appeared tired but not defeated.

"Sit down, Bill, How are Kathryn and Billy?"

"Both are fine, we moved to the Valley. More room to roam around. How is Mary Olive?"

"Nervous, fidgety. She helps me out at the radio station part time. We keep moving around. This infernal Corporation Commissioner is driving me nuts! He doesn't like my ads. He doesn't like me. He is a lackey for the big oil boys. They hate me!"

"Off the record, C.C., was the New York Times article accurate when you were quoted as telling a cab driver, 'You be the fare, I'll drive.'? It then described a wild trip through signals, over curbs, ending in a crash into an automat, costing you $25,000?"

"I was working off my frustration with the bankers."

"Now on the record C.C., were you successful in selling stock during your recent stay in New York?"

"For the record, No! The Easterners say the oil market is stale in California. They say Union Oil is taking wells off the line. Too much oil. I blame my failure to sell Julian shares on my enemies here in Los Angeles, the oil respectables and their banker friends and you can print that! Off the record Bill, yesterday Jim Beebe set up a luncheon appointment for me over at the Biltmore. A stock broker named George Johnson introduced me and one of my business associates to Senator William H. King of Utah who represented to me that he was the attorney for a Mr. S.C. Lewis, an oil man with connections in Oklahoma. Lewis introduced himself. He claimed he was interested in either leasing or acquiring Julian Petroleum Corporation. Bill, will you see if you can find out something about these fellows? I never heard of them."

"OK, Chauncey, I'll look into it for you. S.C. Lewis, I've never heard of, but I believe Senator King is a friend of William Gibbs McAdoo and also has ties with former Senator Frank Putnam Flint. I'll call you on that C.C.. Nothing in print."

Jim Beebe flung open the door and entered Julian's office without knocking. "Mr. Julian, I received a call from a friend of mine over at the Pacific Coast Stock Exchange. The word is out that the Commissioner Dougherty is going to demand our books for inspection again."

"Jim pack up the books. Take two staff. Get the books on the train

< 49 >

tonight. I'll book all of you one way to Wilmington, Delaware away from that snake's reach. I'm going to call a meeting of my stock holders. They will back me. Bill, where would be a good place for a mass meeting of Julian stockholders?"

"C.C., how many stockholders do you have?"

"At least 40,000."

"Well, let's say that at least half of them trust you. Why not rent the Hollywood Bowl?"

"That's it, we'll do it. Jim, get some cash from the vault and you go over and rent the Hollywood Bowl for next Wednesday evening. I'll show these people a thing or two. Also, Jim, arrange for those folks we know out in Rancho Cucamonga to supply us with enough lemons for lemonade for 20,000 people. Pay for it in advance. I want that lemonade available for the stockholders at the Hollywood Bowl. Do you understand?"

"Yes sir, Mr. Julian."

"Chauncey, I'll be covering your speech at the bowl."

"Bill, so will the other papers. Do you know the free press is threatening not to print my ads. After all the money they made off of me. I tell you Bill, if my loyal stockholders, the little guys and I can't salvage Julian Petroleum, I'll take everyone else in this city who hates me down with me. I came in here on that train with you, broke. I made a few bucks here, but I was never accepted. Chauncey Julian is a nasty name in the corporate board rooms and bank board rooms. These guys control the phoney politicians and they don't like me either. Also, I smell a rat in this S.C. Lewis and his big shot attorney. He and his Utah Senator want to use old C.C.. If my stockholders and I can't make it, then we'll see about S.C. Lewis. I don't trust him, but I'll use him if I have to and when I take the last train out of this town, I'll make sure I'm never broke again."

"How about Mary Olive and the girls?"

"Bill, I've kept them out of this mess. If I have to leave town, I'll send her and the children to Canada. I don't want them hurt by any of these boobs."

They arrived in autos, trolleys, and on foot. The Julian stockholders freely checked through the turnstiles of the Hollywood Bowl and were greeted by Julian staff members who dispensed free lemonade from refreshment stands.

"How's the crowd, Jim?"

"Packed to the rafters, Mr. Julian. Well there are no rafters, just the afterglow of the sunset. Do you think we're going to make it, Mr. Julian?"

"This is my night. Let's go get em."

Julian stepped onto the stage in the bright lights of the Hollywood Bowl. The crowd hushed. There was no applause from the 20,000 fearful, yet still loyal stockholders of Julian Petroleum.

"Ladies and gentlemen. Fellow stockholders. Some of you out there were with us at the Glenn Ranch after the ribbon cutting ceremony of our new gasoline station in Rancho Cucamonga. Remember that event? Remember the inscription on the door of the ranch house? They have a motto

< 50 >

out there, "Welcome Guests and Friends." To you folks, I say earnestly, Welcome Guests and Friends! I want to share my concern, which is also your concern, for Julian Petroleum Corporation. Julian Petroleum is a real oil company. It's ours. We have tank farms. We have a distribution system. We have filling stations. Pipelines. What's wrong with Julian Oil Company? There's nothing wrong. The big oil companies are the ones that are wrong. Not only the big oil companies, but the big bankers here in Los Angeles. They are wrong. They have conspired to ruin us. Our wells are producing, but not producing enough to supply us with the crude oil we need to provide the gasoline for all of our filling stations. These big oil companies refuse to sell me crude oil. They have forced me to go around to automobile dealers and gas stations buying up their old crank case oil which we send to our refineries to try to keep going. But it's not enough. The bankers here won't lend us the money we need. I just returned from New York where I spent $200,000 of my own money trying to get some financing, but the Doheny Oil and the Pan American Oil big shots here in Los Angeles cut the ground under my feet by vicious phone calls and letters to the bankers.

That's not all folks, here in Los Angeles, ruthless stockbrokers and fourflushers have beaten down the market value of our stock by forming a pool to control our stock price so that I couldn't get any financing. But I have a plan. I have a plan folks, but you must trust me as you have in the past. I have a plan to give us some time to get our stock up to the $50 a share mark. And here is what I'm recommending to you stockholders today. I'm recommending that all of us, each one of us escrow our Julian preferred stock and then buy up all the cheap stock that these broker scoundrels have swindled us out of by urging you to trade it for mining stock and other worthless securities. My attorney has worked out a plan. We will be getting back to you in writing. Now, how about it folks. All of you in favor of forming an escrow for our preferred stock! Say Aye."

First the stockholders were speechless. Some began speaking one with the other. "Ayes" began popping up here and there among the crowd. Soon a ground swell of "Ayes" led Chauncey Julian to raise his arms to the sky and proclaim "Julian Petroleum Company will be saved!" Julian left the stage and the stockholders slowly drifted from the area. A group of newspaper reporters and press photographers gathered near Jim Beebe. Among the group, was Bill Feeney accompanied by Raul Dominguez, camera in hand.

"Mr. Julian," a reporter from the Hearst Paper; "Mr. Julian.."
"Yes sir."
"How much of your own money have you invested in Julian Petroleum apart from possibly stock option purchases?"
"Sir, I have advanced slightly over one million dollars which I earned on commissions. Besides my concern for my 40,000 stockholders, I have a personal commitment to salvage Julian Petroleum from the ravages of government bureaucrats, ruthless bankers, and selfish oil barons."
"Mr. Julian, Murphy from the Los Angeles Times. If you had nothing to hide, why did you take your books to Wilmington, Delaware to avoid inspection by the Corporation Commissioner?"

< 51 >

"We are a Delaware Corporation. I am fed up to here with snooping bureaucrats. This is a free society. I don't recall the Corporation Commissioner snooping into Union Oil Company's affairs when the Europeans were trying to take over their precious California Corporation. These politicians like to pick on little guys. The politicians need the money the big oil people give them and they get left alone."

Bill Feeney whispered to Raul Dominguez "Raul, get a shot of Julian when he raises his arms."

"Mr. Julian, Feeney from the Hollywood Citizen News. Earlier this afternoon, just prior to your speech, Associated Press placed on the wire service a story from Sacramento. It stated that the state Corporation Commissioner had impounded $1,800,000.00 of Julian Petroleum funds and has brought charges against you for illegal stock selling, deception, manipulation and evasion. Have you any comments Mr. Julian?"

"This is the first I've heard of these charges."

Chauncey Julian raised both arms upward and shouted, "I will never see one minute in court based on these phoney trumped up charges. I will have no further comment until I see my lawyers."

"Mr. Julian, Jones from the Long Beach Press Telegram. Today our paper has announced it will no longer print your ads. Any comment?"

"They sure took my money for the past two years. I'll tell you this, Wrigley is selling his chewing gum on 573 radio stations in the United States and not paying a dime to a newspaper. I can do the same thing. I'll be using my radio station to advertise my stock offerings in the future. Anything else gentlemen, if not, I thank you and I bid you goodnight."

"Raul, did you get that shot of Julian with his hands outstretched?"

"Sure did, Bill. They may be in handcuffs if the Corporation Commissioner is serious."

"Excuse me, Mr. Feeney. May I have a word with you?" Postal Inspector Madeira flashed his badge attached to a leather case containing his photo and identification card.

"Excuse me, Raul. I'll be back in a few minutes. Feeney and the Inspector moved away from the reporters.

"Feeney, I've obtained copies of photographs. You appear in several taken in Cucamonga during the dedication of one of Julian's gasoline stations. I also observed you and your wife and child in your Auburn with Mrs. Julian. Do you have some special relationship with Julian?"

"I don't think it is any of your business, Inspector. I arrived in Los Angeles with Julian on the same train, purely by chance. We have kept in touch. I'm no insider. I'm a professional. What are you after?"

"Nothing in particular, Feeney, just curious. I'll see you around."

CHAPTER 7

"Bill, turn around. No, don't start that again. I put on my negligee. Don't you ever wear your pajama top?"

"Not since I met you. Come here."

< 52 >

Bill rolled to his side, pulled back the bed sheet and blanket, and gently pulled downward on Kathryn's black gown exposing her breasts which he kissed ardently one, then the other. He kissed her lips with strength and affection and then he gently replaced her bodice, sensing her reluctance.

"Bill, thank you again for this new home. You know how much I love you. You don't tell me you love me very often, but you do things to show your love. There's something on my mind. Do you think I should work full time over at First National since we live so near the studio?"

Without answering immediately, Bill switched on the night stand lamp checking his watch on a dressing table. It was 6:00 am.

"Kathryn, Billy will be two in February and this will be our first Christmas in our own home. I would really prefer you spend your time with the baby. I'm making extra money from my articles for Sunset. But, go ahead and establish yourself over at the studios. Try to time your work so most of your time will be with the baby and me. When Bill starts school we can take a look at both of our careers."

"All right. Taking care of a baby is a lot of work. By the way, you'd better get an ample supply of your illegal French condoms. I don't want to get pregnant again for at least six years."

"Yes, I play a little game with the druggist. I tell him I don't want to get venereal disease and he sells them to me. I'm ready for my morning run around the lake. How about some breakfast when I conme back, Canadian bacon and eggs?"

"Speaking of Canadian Bacon, I'm thinking of inviting the wife of that Canadian ham Julian over for tea before Christmas. She called the other day. Julian is out of town constantly. She claims their daughters hardly know they have a father."

"That's all right with me. Just don't put the squeeze on her for any information about her husband or Julian Petroleum. I have to keep my distance from that fellow and maintain a professional approach. Postal Inspector Madeira is on his case. Julian is running a bit on the wild side as your Hollywood cowboys call it, Kathryn. Where are my sneakers and sweatshirt? They're not in the closet?"

"The sneakers are on the back porch. You have clean socks and a clean sweatshirt in your second drawer. Just look for them. And please leave your wet sneakers on the back porch! Kiss me goodbye, I love you. What time do you want breakfast?"

"Six forty-five. I have to be at the office by eight. I'm meeting Raul for an assignment to interview the Junior Senator from the state of Utah, the Honorable William H. King, Democrat, of Salt Lake City. He's having a press meeting at the Biltmore Hotel. Hank Bond says King is the hardest worker in the Senate. He wants to address the local press for some reason. Strange to say, C.C. Julian dropped his name the other day. He wanted me to get a line on Senator King, but I'm staying away from dealing with Julian on anything about King. As far as I know, the only thing those two have in common is a reputation for being hounds for work. See you in a little while."

Bill found his sneakers on the back porch. He put on a cap, and tucked a woolen scarf inside his sweatshirt. He closed the back door and

< 53 >

jogged toward Toluca Lake.

The couple had combined their savings and purchased a lot from a Hollywood real estate agent in a new tract development near Toluca Lake, a section of the town of Lankershim, located in the eastern portion of the San Fernando Valley. Borrowing from a bank, they financed the construction of a Spanish style, stucco home, with red tile roofing similar to the flat they had occupied in Hollywood at the time of their marriage. The house was small and lonely, the first one built on Forman near Valley Spring Lane about a quarter of a mile from Toluca Lake.

Bill ran toward the lake. Dawn was approaching. Camel's Back Mountain, separating the edge of the San Fernando Valley from the Cahuenga Pass and Hollywood was etched in the early glow of the winter sunrise. Cool breezes off the lake energized Bill Feeney for a fast run.

Returniong home, Bill showered, shaved, dressed, scanned the morning paper during his breakfast, kissed his wife and the baby good-bye, and headed off toward the Cahuenga Pass for a twenty minute drive to the Hollywood Citizen News.

Kathryn changed the baby's diaper, put him in his playpen, and placed a telephone call. "Operator. Granite 3414 please." The phone rang several times.

"KMTR Radio, Mary Olive speaking may I help you?"

"Mary Olive, this is Kathryn Feeney. How are you?"

"How nice you called, I am bored again but keep working at the station to keep busy. The children are in school and I come here just to pass the time. I leave the station when Chauncey arrives for his broadcast. He built an enormous home in the Las Feliz area. I get bored just sitting around there. I have no friends. Chauncey is always away at work or doing something."

"Listen, Mary Olive, I would love to have you come over and see our new home. Can you pop away for tea or lunch?"

"Today is about the only chance. Chauncey is due back in town tomorrow. A lot of things are happening."

"Great, Mary Olive, I'll meet you at the main gate of First National Studios. Our home is in the middle of nowhere. You would never be able to find it. Will you be driving?"

"Chauncey's Pierce Arrow. He lets me drive it when he's out of town. Is noon too soon?"

"No perfect. I'll be in my Ford. I'll honk when I see you and then you follow me."

"Should I bring anything?"

"No, we have a little store here. We're not too primitive. I'll see you at noon. Bye for now."

Kathryn changed the baby's clothes, placed him in the baby seat, and drove to the Weddington General Store in Toluca to shop for the luncheon and for her husband's dinner.

Mary Olive Julian had no official position at the radio station. She served as a sort of receptionist at KMTR. Jim Beebe, the General Manager of the station, hired staff and was responsible for preparing the recordings,

< 54 >

which he called Hollywood hillbillies as Julian wanted him to play Western style music. The studio was located in an office building on Hollywood Boulevard near Wilcox. The station's antennae, two steel towers, were located on the roof of the building. The reception office which Mary Olive occupied, contained a telephone switchboard and was in a glass enclosed cubicle with a second floor view of Hollywood Boulevard.

In the studio, heavily curtained with thick green drapes, on all four walls, stood transmission equipment, a desk for the engineer, and a table which contained the announcer's microphone and turntables for playing recordings. The announcer, also had a telephone available to answer questions or communicate with the receptionist when needed. If a telephone call came in for the announcer the receptionist switched on a red light. The station played country western music from six in the morning until six in the evening. Julian purchased the station from a Hollywood automobile dealer who in 1922 had applied to the United States Secretary of Commerce for a license and frequency.

"Jim Beebe, here."

"Jim, this is Mary Olive. Please have someone come over to handle the telephone switchboard at the station. I was invited out to lunch and I'd like to go."

"Certainly, Mary Olive, I'll be over by 11:00. I was going there anyway to bring a new batch of recordings."

"Thank you, Jim. Have you heard from my husband?"

"Yes, he called from Oklahoma, he'll be in town tomorrow."

"Thank you, Jim."

Mary Olive drove through the Cahuenga Pass to Dark Canyon. She turned right on the narrow road for approximately half a mile with the road then descending into the San Fernando Valley. She crossed a wooden bridge which spanned the Los Angeles River. Brownish colored water flowed eastward through green willows. Below the bridge children were picking watercress. She arrived at the main gate of the First National Studios. Kathryn was waiting. Mary Olive waived and followed the Ford through a subdivision of empty lots with FOR SALE signs giving a Hollywood phone number of GRANITE-4411. Turning left on Foreman, the autos descended parking near the intersection of Valley Spring Lane. Mary Olive parked her car beneath a sign warning, "KEEP HORSES OFF SIDEWALKS." Kathryn took Mary Olive's hand, helped her out of the car, hugged her, and retrieved the baby from the car seat. In the entryway, Kathryn excused herself for a moment, and placed the baby in his playpen near the breakfast nook off the kitchen. She helped Mary Olive remove her coat, placing it in a closet off the entryway. Mary Olive felt at ease. The Feeney's had acquired a complete set of furnishings, the overstuffed type, comfortable, but not gaudy.

"Mary Olive, come to the kitchen while I fix lunch. I have to keep an eye on the baby."

Mary Olive followed Kathryn through the small dining room to the breakfast nook and sat near the baby's playpen.

"Do you enjoy your new home, Mary Olive?"

"Not really, it's all Chauncey's idea. It's much too big. He has

< 55 >

dreadful taste in decorating."

Kathryn busied herself with the final stages of a meat loaf, whipped potatoes and a tossed green salad.

"Do you prefer tea or coffee, Mary Olive?"

"Tea, please, Kathryn. Kathryn, are you doing anything at the studio?"

"Not much. A few bits here and there. Bill wants me to spend time with him and the baby. How about you, do you enjoy your work at the radio station?"

"Yes, but I only work there occasionally and basically to kill time. Chauncey is very proud of his radio station. He says it gives him more freedom to say what he wants to say without interference from newspaper editors. He has his own broadcast each week night from 6:00 p.m. to 6:15 p.m. Chauncey says the radio station allows him to keep in touch with his stockholders. The remaining time we supply recorded musical programs. Do you ever listen? Western style, the Hollywood hillbillies. Also, some companies pay us to advertise their products. Chauncey is working on a deal with that chewing gum company.

I'm happy your husband provided you with this home. Chauncey is only interested in his business. He makes an enormous amount of money, spends it foolishly, like that big house for example. He's thinking about having the children and I return to Canada, if things don't work out for Julian Petroleum. I don't know, I get so lonely. I'm sick and tired of the way the children and I are living. I never meet any family people. I thought I would meet a lot of people at the radio station. Sometimes Chauncey asks me to come to the station and answer the phone after one of his speeches. But all I hear are voices. I don't really know any of the people. I don't think I'll ever get used to living here."

"Mary Olive, almost everyone in this town is from somewhere else. Lots of us, including my husband, came here partly because being strangers gave us some freedom to develop our own careers without family interference. Mary Olive, for today let's forget loneliness. It's almost Christmas. We have our friendship as long as you are here in Southern California. Forget about our husbands. They have their jobs. You need to concentrate on your children. Tell your husband you don't want to work at the radio station any longer."

"That would be worse, Kathryn. I would go crazy sitting around that big home all day and all night with Chauncey gone so often."
Kathryn fed the baby. "Is there a school nearby for Bill, Jr.?"

"Yes, Mary Olive, there are two schools in Lankershim. By the time he's old enough, I'm sure they will have a high school. There is no hospital out here, but it's not too far from Hollywood Presbyterian where the baby was born."

"Are you going to have more children?"

"Not for at least six years. Taking care of one child is full time work. I miss my career and as soon as this child is in kindergarten, I'm going to work full time."

Kathryn picked up the baby and placed him in a high chair near the dining room table. She served lunch in the dining room.

"Kathryn, does your husband philander?"'

"I don't think so. He goes to work early. He comes home early. He works a bit on magazine articles here at home on his typewriter. I never thought of him as a philanderer. We have a good sex life. Why do you ask about philandering?"

"Because Chauncey is a philanderer. I've asked him to stop it, but I know he won't. This big home is supposed to satisfy me, but it's drowning me. I'm miserable."

"Please, Mary Olive, just for today, let's forget about our husbands. But if Bill was ever unfaithful to me, I would take this home, the child, and I would make it known to the whole world what a rotten fool he was to destroy our family. Oh, the phone is ringing. Excuse me for a moment."

"Hello, this is the Feeney residence."

"Hello Kathryn, this is C.C.. I understand my wife is having lunch with you."

"Yes she is, are you checking up on her?"

"Exactly. She's been acting in a peculiar manner lately. Did she say anything bad about me?"

"C.C., I think you'd better talk to your wife. Mary Olive, your husband would like to speak with you."

"Hello Chauncey, how are you?"

"I'm just fine. I'm staying for a few days in my apartment in Oklahoma City. I'll be home the first of next week."

"How is Miss Smith, Chauncey?"

"Miss Smith is just fine, my dear. Say hello to the girls. I'll see you next week. Good-bye."

Chauncey Julian replaced the white telephone on the marble table bed stand. June Smith threw the pink down comforter aside. She sat upright in bed supported by five pink silk covered pillows. She was smoking a cigarette. "Chauncey when are you going to get rid of that bitch?"

"Be patient, my dear. I have a plan. But it will take some time. She seems to be developing a mental problem. I may have to do something about the children. Tell the folks in Texas anything you want to about our marriage plans. It's just a matter of time. Honey, I'm building a swimming pool in the shape of a heart just for you."

"Oh, Chauncey, can I tell daddy and mommy?"

"Yes, June honey, tell them the pool in Hollywood is just for you."

"This is Postal Inspector Manny Madeira. Please connect me with David Cahill. Thanks."

"Dave, this is Manny Madeira. Is it snowing in Washington?"

"No, but it's cold! What's up?"

"Julian is in Oklahoma City with his girlfriend. Beebe is running back and forth between the radio studio and Julian's downtown office. Today, I followed Mary Olive Julian. She drove Julian's car to the Valley. She met that newspaper reporter's wife near the main gate at the First National

< 57 >

Studios in Burbank. She then followed her to a home near Toluca Lake. Mary Olive was in tears when she left the place. Darned near crashed her car. She seemed frantic. I still don't understand the connection between Feeney, his wife and Julian. I'll keep you posted. So long."

"So long, Manny."

CHAPTER 8

"Where to, Mr Julian?"

"Take me home, Jim, Hollywood, Mary Olive. I'll have to listen to the old cat's tonsils."

Chauncey Julian handed Jim Beebe one of his two leather grips. Julian exited the train from Oklahoma City at the East Los Angeles station. He hoped to avoid any stray stockholders who might buttonhole him at the Los Angeles station.

"June is staying in Oklahoma until after New Years."

Jim Beebe placed the two leather grips in the trunk of the Pierce Arrow. Chauncey Julian got in the back. Beebe drove westward on Whittier Boulevard toward downtown Los Angeles.

"Jim, I've decided to appoint Sheridan Lewis General Manager of Julian Petroleum. He has a pussy foot approach to the big shot bankers. He has buffaloed them with his so-called attorney — Senator King and King's friend William Gibbs McAdoo. I told Lewis the stockholders would be shocked if I moved out of the company too quickly. I asked him for one million dollars which I loaned the company. King looked at an audit and agreed to give me $500,000 cash for the company. We will keep one of our offices at the Loews State Building. Lewis will be opening offices near the Biltmore Hotel where he's living. Ship the corporate books to our New York office. Take care of that right away. My board and I will resign when I get the cash. I want you to stay with me at the radio station. Do you understand?"

"Yes, Mr. Julian."

"Also, I don't want Mary Olive hanging around the radio station any more. I'll talk to her about it tonight. While Lewis boot licks the bankers, they will move in with their own board of directors. The bankers promised to make the necessary loans to satisfy the stockholders. Then I'll announce I'm out of the picture. I'll tell everyone that leaving Julian Petroleum was the worst boob trick I ever pulled. After the first of the year we will get started selling stock in Western Lead Company. The newspapers will print my ads because they won't know what I'm up to. We'll use the radio station to help promote Julian Petroleum for the time being. Lewis has the bankers and politicians working like sixteen oiled cylinders. The only advice I gave Lewis about his politician friends is to wiggle in a new governor at the next election. That's the only way he can get rid of that son of a biscuit eater corporation commissioner Dougherty. Lewis will probably try some of his oily scheming on Governor Richardson, but I don't think it will work. Jim, when we get to Hollywood and Vermont, stop at the Presbyterian Hospital and have a cab

< 58 >

follow us home. You can take the cab home. I may need the Pierce Arrow. When June arrives here after the first of the year I'll trade the Pierce Arrow in on a Rolls Royce I ordered."

Jim Beebe drove north to Hollywood Boulevard, turned left and after a few short blocks turned right on Edgemont. On a gentle slope toward the base of the Hollywood Hills, at Cromwell Place, Beebe parked the car in front of a locked wrought iron gate. A gentle rain that began in the early evening slackened to a drizzle near midnight.

"Driver, wait for me here."

The taxi driver nodded as Julian gave Beebe the key to the padlocked gate. Beebe drove up the sloping slick concrete driveway and parked near the swimming pool and red brick barbecue pit. He helped Julian with his luggage to the rear door of the mansion.

"Goodnight, Mr. Julian."

"Goodnight, Jim. Here's $500 for your help. I'll see you at the radio station sometime next week."

Beebe returned to the main gate and secured it with the padlock. The taxi driver turned his cab to head down the hill.

"Driver take a left turn on Cromwell Place."

The Julian mansion stood on a promontory at the corner of Edgemont and Cromwell Place. Massive, constructed of burnt red brick in an architectural Baronial style, its four and one half acres were surrounded by a seven foot brick wall. The closed wrought iron gate secured the main driveway while a secondary smaller entrance for servants and guests at outdoor parties was secured by a gate near the eastern portion of the property. The rain slickened shake roof was illuminated by several flood lights, silhouetting the mansion and its tall brick chimneys. As Jim Beebe's taxi cab descended in the mist toward the lights of the city below, Beebe caught sight of a lighted Christmas tree near the leaded glass window of the living room and the silhouette of a woman seated near the window of the library, the only room in the house obviously occupied at the midnight hour.

Julian entered the back door using a key hidden in a recess of a built-in milk delivery box on the wall near the back porch. Carrying his two leather grips, He made his way to a dark entry hall. A slit of light shone brightly at the bottom of the door leading to the library. Julian dropped his suitcases noisily to announce his presence and slowly opened the library door and stood silent and motionless in the doorway. The back of a woman's head protruded slightly above a chair which faced the window. A hand moved outward from the chair placing a book on an end table.

"Courtney."

"Mary Olive."

"Courtney, how thoughtful of you to stop by."

"Mary Olive, I didn't come here to. . ."

"To what, sleep? Speaking of sleep, is June with you?"
Julian entered the library and closed the door. "No." "By the way, Courtney, why do you carry valises to Oklahoma when your apartment is fully equipped? Of course dear Courtney, for someone who has everything there is yet the issue of propriety, is there not? One needs to give the right

< 59 >

impression. You always do, don't you, Courtney?"

"You ungrateful bitch, don't talk to me like that. You sit here in the largest home in Hollywood, the most elegant home in Hollywood, surrounded day in and day out, in luxury. You have the gall to lecture me? You would still be sitting up there smelling the hog manure in Manitoba if I hadn't brought you and the girls down here"

"Courtney, I'm not looking at you, but I know what you are, not what you appear to be. You are the same old plumber's helper I was fool enough to marry. Now you're richer but you still have no class. That's why important people stay away from you. They see through you. This is probably the first time you were in a library more than two minutes in your whole life."

"I may not have much class Mary Olive, but I do know how to pick women, how to handle them, and how to keep them in their place. Every one I've ever picked is more beautiful than you."

"More beautiful, Courtney? Is that a slip of the tongue or is your grammar poorer than usual?"

Julian walked to a telephone at an oak desk. He picked up the receiver, placed it to his ear and turned and faced Mary Olive. Julian gave the operator a GRANITE phone number.

"Winnie, darling, this is Chauncey. I'm at the train station. The damn train was delayed by a desert storm. I'll be over in half an hour. See you then."

Julian then puckered an audible kiss into the phone receiver and hung up. Julian placed a second telephone call still facing Mary Olive.

"Yellow Cab, send a car to the Julian residence immediately. Thanks."

"Goodnight Mary Olive."

"Aren't you going to ask about the girls, Courtney?"

"I'll call the girls in the morning from the office.
And quit calling me Courtney."

"I can't call you C.C. because your friends all call you C.C.. I can't call you Chauncey because your girlfriends call you Chauncey. Goodnight, Courtney."

Julian closed the door to the library, turned on a light in the entry hall, chose one of his canes from the closet along with a derby hat and an overcoat and waited for the lights of the taxi cab.

"Driver, take me to the Kingsley Tower Apartments on Franklin Avenue."

"Yes sir."

Julian gave the driver a fifty dollar bill from a wad of fifties and hundreds. He entered the luxurious apartment tower, pushing the elevator button for the seventh floor. At the top floor of the tower, he removed a key from his wallet which activated a second smaller elevator. The mirrored elevator quickly ascended to the penthouse. The door opened automatically to a living room decorated entirely in black and white. A floor to ceiling window provided a view of Hollywood Boulevard to the south.

"Winnie, darling."

< 60 >

A tall flaxen haired woman stood near a black and white mother of pearl oriental screen which partially hid from view a door to the penthouse bedroom. The woman smiled at Julian. Julian placed his derby hat and cane on a glass table. He removed his overcoat and tossed it on the white sofa. He extended h is arms and approached the woman who stood motionless with her arms at her side. Beneath a white silk robe she was wearing a pink negligee.

"Let me look at you."

Julian gently grasped the woman's hands and kissed her. "I've missed you, Chauncey."

"I've missed you, too, Winnie. You look lovely as always. Let's go see our cowboy friend."

Chauncey Julian escorted Winnie St. Cyr to a dining room off the main living room with yet another large window facing the Hollywood Hills.

"Let's see what the cowboy is up to."

Winnie placed her arms on Julian's shoulders. Julian reached into his hip pocket and took out a billfold. One by one he removed ten five hundred dollar bills draping them over a whip held in the hand of a cowboy immortalized in bronze.

"This is for the old landlord. This is for you. This is for the luggage I want you to buy. You and I are going to Europe for a whole year. Just the two of us. We'll go down and get the paperwork moving for passports this week."

Julian led her to the couch where he had tossed his overcoat.

"Winnie, honey, hang up my overcoat. Before you do, reach into the right hand pocket and see what you find."

Winnie picked up the overcoat, searched the right hand pocket, withdrew a small black box, tossed the coat on the sofa, and opened the box. "Oh, Chauncey, you know how I love emeralds."

She turned toward Julian surrendering to his open arms. She placed the emerald ring on her right hand ring finger.

"Chauncey, I love you."

Nine am. Julian was awakened by a ringing of the telephone on the dresser. He shook his head, slipped into a white silk dressing gown and answered the phone.

"Hello."

The voice on the other end was Jim Beebe.

"Mr Julian, I have some bad news."

"How did you get my phone number?"

"I called Mary Olive. She told me she has all of your numbers memorized."

"All right, what do you want?"

"Federal agents met me at the office door this morning. They have a search warrant. They want me to open the safe. They want our records."

"Who's in charge, Jim?"

"Postal Inspector Madeira and a Treasury agent."

< 61 >

"Let me talk to Madeira."

"Madeira, here."

"What are you up to, Manny?"

"You're not paying Federal Income Taxes compared with the specific items you own or lease. For starters, your home in Hollywood, three apartments for your girlfriends, the new Rolls Royce. I have a court order to seize your books and seal your safe deposit boxes."

"Does your court order limit you to the Julian Petroleum Books?"

"Yes, it does."

"Then don't touch any documents relating to my mining projects."

"We have no interest in your mining projects."

"Let me speak to Jim Beebe. . ."

"What shall I do?"

"Open the safe."

"What else?"

"For the time being do nothing else. The holidays are coming. Federal agents are not like cops. They take holidays just like we do. After the first of the year we'll see what they try to do. Set up a meeting at our offices Thursday after New Years. I want both my lawyers present, my tax attorney and the one that deals with tax problems at Julian Petroleum. Winnie and I are going to be busy this week so don't bother me anymore. We will be staying at Catalina until after New Years. If June Smith arrives in town, tell her I'm out of town and will call her when I get back. She'll be staying at her apartment on Berendo Street. I won't be doing any broadcasts at the station for a couple of weeks. Use records of old broadcasts. Jim, is Madeira seizing any documents about the mining operation?"

"No, he looked at the corporate papers but didn't take them. They're still in the safe."

"Listen, Jim, take the corporate papers over to our printer and have him make up ten thousand stock certificates. He has a machine that prints my signature. As soon as the holidays are over, I'm going to organize a campaign to sell the lead mining stock. Take care of that will you, Jim?"

"I sure will, Mr. Julian."

Julian turned to Winnie who was seated at her dressing table admiring her emerald ring. He leaned over, staring at his mirrored image next to hers. Gently massaging her shoulders, he kissed her on her right cheek. She snuggled up to him.

"What was the phone call about dearest?"

"Nothing, just the tax people. I'll take care of it. Let's go over to Musso & Frank's Grill for breakfast. We need to start making our plans for Europe."

CHAPTER 9

"The Tournament of Roses made a wise choice." Bill Feeney called to Raul Dominguez in the tumult at the Central Railroad Station as Fay Lanphier, under the chaperonage of her mother, Mrs. Emily Lanphier, was

< 62 >

greeted by members of the tournament association. Representatives of the press, local, national, news services, and news reels crowded near the reception area as America's prize peach, the winner of the 1925 Miss America Contest at Atlantic City was welcomed as Queen of the 1926 New Year's Day Tournament of Roses. Dressed in a long fur coat and tight fitting knit hat, Miss Lanphier stepped down from the Pullman car and was handed a bouquet of roses. A welcome was extended by Mrs. R.I. Strand, the daughter of the director of the Tournament Association.

Reporters were asked to withhold questions until after a luncheon which was scheduled at the Valley Hunt Club in Pasadena. Bill Feeney took notes, while Raul Dominguez struggled through the ranks of the other press photographers to get a close shot of the beauty from Oakland.

As the hubbub subsided, the queen and her mother were escorted to a limousine and whisked away behind uniformed police on motorcycles. Bill Feeney completed his note taking. Apart from the crowd, he noticed a familiar face. Chauncey Julian stood near a baggage car which immediately preceded the rose queen's Pullman car. Julian was with a woman who paced back and forth as Julian involved himself with some sort of paperwork.

"Raul, let's stop and say hello to Julian. He's been in Europe for almost a year."

Feeney walked away from the crowd of reporters toward Julian and his companion. Julian was supervising the unloading of crates from the baggage car. Dominguez followed at a slower pace, stopping to change the frame in his camera.

"C.C., welcome home."

Julian turned toward Feeney neither elated at the welcome nor displeased.

"Hi Bill." The two shook hands.

"Bill, this is Miss St. Cyr. She accompanied me to Europe as an adviser. We purchased art works and rare books. We have cases full of things I've just cleared through New York Customs. It cost me a pretty penny."

"Pleased to meet you, Miss St. Cyr."

She extended her hand which Bill shook gently. She wore a ring with a large precious stone. Rested and displaying none of the tension and anger of the previous year before stepping down as President of Julian Petroleum Corporation, Julian seemed carefree and poised.

"Miss St. Cyr is a college graduate — also trained in nursing. She makes sure my old ticker is on time. Are you working at the Rose Parade tomorrow?"

"Yes I am, C.C., we just met the Rose Queen. Are you going to the football game?"

"Probably, Jim Beebe has some tickets for us. I'm taking Miss St. Cyr to the parade. We have reserved seats in a reviewing stand. By the way, has your wife visited Mary Olive lately?"

"I believe so, they have lunch about once a month. Woman's talk, I guess. Listen C.C., I have to get my story ready for the afternoon paper. I'll stop by your office next week and talk a little more about Julian Petroleum and the new management."

< 63 >

"Anytime, Bill."

Feeney walked over to a baggage truck where Raul Dominguez had placed his camera.

"What did Julian say, she takes care of the old prigger?"

"Ticker, Raul. Did you see the rock on her finger? She must be a valuable connoisseur of objects d'art. Let's get back to the paper."

Chauncey Julian continued supervising the transfer of a dozen or so cartons and suitcases from the train baggage car to an express delivery truck, checking items on a lading slip. Winnie St. Cyr stood by patiently. Shortly Jim Beebe arrived, driving his employer's Rolls Royce.

"Good morning, Mr. Julian, nice to have you back. You, too, Miss St. Cyr."

"Hi, Jim, I think things are in order. Are our reservations in place at the Vista Del Arroyo Hotel?"

"Yes, Sir."

"How about the reserved seats for the Tournament of Roses Parade?"

"Yes Sir, your tickets will be with the concierge at the hotel. We also obtained two tickets for the football game in case you are interested. Do you want me to pick you up at the hotel and take you to the parade tomorrow?"

"No, we'll take a cab. Meet me at the hotel at about 4:30. If I go to the game, I'll take a cab, if I don't I'll see you at the hotel. We'll be returning to Hollywood after the festivities."

"Winnie, dear, please get in."

She entered the rear seat. Chauncey Julian leaned inward.

"Just a minute, Winnie, I forgot something."

Julian signaled Jim Beebe back to the baggage car.

"Have you heard anything from June? Is she in town?"

"I don't think so Mr. Julian, she hasn't called the office."

"How about Mary Olive, has she bothered you much?"

"Not too much, she calls about once a week, claims you don't provide her with enough money to run the home. I send her twenty five hundred a month like you told me."

"That's all right, I'll stop by there tomorrow evening to visit the girls. After you drop us off in Pasadena, stop by the home in Hollywood and make sure the crates on this lading list are stored properly in the guest house for the time being. Now let's go."

Chauncey Julian entered the rear seat of the Rolls Royce and Jim Beebe headed for Pasadena, driving east over the Macy Street viaduct. Chauncey Julian held Winnie St. Cyr's hand.

"What's the weather forecast for tomorrow, Jim?"

"The paper predicted generally fair, Mr. Julian."

New Years Day, 1926. Chauncey Julian and Winnie St. Cyr took a cab from the Vista Del Arroyo Hotel to the corner of Madison and Colorado Boulevard in Pasadena. Julian surrendered their reserved tickets for row thirteen of a twenty tier review stand. Julian and Winnie joined the crowd.

< 64 >

The parade was scheduled to begin at 10:30 a.m.

The first contingent arrived in front of the reviewing stand. A solo bugler preceded the Pasadena Elks Band. Three official cars carried Lieutenant Colonel L. J. Myatt of South Pasadena, the Grand Marshall, Mayor Cryer of Los Angeles, followed by an official car with Mayor Rolph of San Francisco, who had been honored with an invitation to ride in the parade. This was the first time in Rose Parade history that a float representing the City of San Francisco was featured.

Following the official cars carrying the dignitaries came a series of carriages depicting the tournaments of earlier years, including a horse drawn carriage celebrating The Old Valley Hunt Club, the organization which founded the tournament, a horse drawn fire apparatus, and men riding old fashioned high-wheeled bicycles.

The float section of the parade was led by the Tournament of Roses Association on which Fay Lanphier, Miss America of Atlantic City's beauty contest rode surrounded by her attendants. As her float passed the reviewing stand, Chauncey Julian and Winnie St. Cyr joined the crowd and noisily cheered and applauded. Miss Lanphier's throne was a composite of roses and carnations, while before her was a bubbling fountain of youth. The float looked like a beautiful ship pulled by a span of white horses.

Suddenly without warning, a sound resembling gusts of wind echoing in a marble mausoleum was heard, as the reviewing stand collapsed like the spreading of an enormous fan. The collapse hurled one thousand men, women, and children in to a mass of struggling humanity. The sound of the collapse was followed by the shrieks and groans of people trapped and entangled with themselves.

An elderly woman standing to the west of the scene of destruction, grasped her chest and collapsed, victim of a heart attack. But Chauncey Julian, Winnie St. Cyr, and others seated in row thirteen slid down the board seat toward the east and were not trapped by the crushed spectators or the splintered wood.

Winnie St. Cyr and Julian helped people to their feet. Members of the Harold Robert's Band sprang into action at their leader's command and placed their instruments on Colorado Boulevard and started to help the injured. Meanwhile, the entire parade stopped while police, sailors and soldiers came to the assistance of the crowd.

Chauncey Julian gathered taxi cabs and private cars for transporation to the local hospital. Winnie St. Cyr gave emergency aid to those with fractures and sprains. A car bearing a parade official drove up and down Colorado Boulevard appealing through a megaphone for doctors and nurses. Waves of cars, taxis, and ambulances transported the injured from Colorado Boulevard to Pasadena Hospital where doctors and nurses treated over two hundred and fifty. Floors were used when beds were filled. Maternity patients in the hospital yielded their beds to the more critically injured. General Hospital in Los Angeles sent every ambulance in their garage, with the except one to assist in the transportation.

As order returned, the parade began once again with many in the huge crowd to the east oblivious as to the magnitude of the disaster. As the

< 65 >

last of the injured were removed, Chauncey Julian, in shirt sleeves, stood next to Winnie St. Cyr, who with other nurses recruited from the parade crowd, assisted in forming a reverse triage of victims.

As the crowd dispersed at the end of the parade, Winnie St. Cyr and Chauncey Julian trudged side by side west, following the movement of people, arriving at the Vista Del Arroyo Hotel shortly after 2:00 p.m. Julian gave away his two tickets to the Rose Bowl Game, Alabama vs. Washington State. He and Winnie were exhausted mentally and physically as a result of their ordeal. In the hotel room, Julian collapsed on a sofa while Winnie bathed and changed into clean clothing. Julian placed a phone call.

"Jim, come over and pick us up at the hotel as soon as you can. There was a tragedy at the parade. Our stand collapsed."

"Yes, I read about it on the wire service. Are you all right?'

"Yes, but we're exhausted and we're not going to the game."

"I'll be there in half an hour."

Jim Beebe arrived at the hotel and escorted Julian and Miss St. Cyr to the Rolls Royce. He dropped off Winnie St. Cyr at her apartment and Julian at his home in Hollywood. Beebe arranged for a cab to follow the Rolls Royce to the Julian mansion.

Chauncey Julian found Mary Olive with his daughters in the living room listening to a musical program on the radio. Julian kissed each of his daughters lightly on their cheeks and turned to his wife.

"Did you get my letters?"

"Yes."

"Have you opened any of the crates in the servant's quarters? I bought some paintings and tapestries. Also a large number of rare books."

"No, I'll leave that to you."

"How is your school work, girls?"

"Just fine, father," the sisters responded in unison.

"Good, good. I bought each of you a jewelry box in Paris. We'll open the crates tomorrow."

"Why is your suit so filthy? You look like a hobo?"

"I went to the Rose Parade. There was a disaster.
Hundreds of people were hurt. I'm all right. Lucky—in row thirteen. We slid down."

"Thank goodness you didn't think to invite your daughters. They might have been hurt more than usual."

"Mary Olive, your thought process is deteriorating. I'll be down for dinner."

Monday, January 4, 1926, Julian returned to his office suite at the Loew's State Building at Seventh and Broadway in downtown Los Angeles. S.C. Lewis, as President of Julian Petroleum Corporation, opened separate offices at the Pershing Square Building, near the Biltmore Hotel. Jim Beebe maintained Julian's office while Julian was in Europe, as well as the KMTR Broadcasting studio in Hollywood. Julian's first move was to place a phone call to S.C. Lewis. It had been over eleven months since the two men

< 66 >

communicated verbally.

"Mr. Lewis please, Mr. Julian calling." After a pause, "Sheridan, this is C.C."

"C.C., how are you?"

"Great. I hear you have been doing well with financing."

"Could be better, could be worse. Listen C.C., I would deeply appreciate it if you could loan us a little of your charisma. You remember Jack Bennett don't you? He previously worked for me in New York. He's a fine young man with a lot of valuable connections. Listen C.C., Jack has put together some pools of Julian stock. We obtained the cooperation of bankers and stock brokers while you were away. But there's only one Chauncey Julian. We need you to pump prime the stock with your loyal followers. Do you think you could place one of your famous ads in the paper for us? I guarantee you they will print the ad. The word is out that the big bankers want Julian Petroleum to go full throttle."

"Sure Sheridan, send Jack Bennett over to my office with ten thousand dollars and I'll plant an ad in all of the dailies on January 11th."

"C.C., just remember our agreement. No more promotions old boy."

"Sheridan my friend, when C.C. Julian tells you he is out of the oil business, he is out. Just send the money over to my office and I'll do the favor for you and your banker friends. I'm glad they appreciate old C.C.'s style."

"Mr. Feeney."

"Yes Miss Carr."

"Mr. Feeney, sorry to disturb you, but I thought you would be interested in some ad copy submitted by Chauncey Julian. Mr. Beebe delivered it on his behalf representing Julian Petroleum. He paid cash."

"Thanks, Miss Carr, let me see it."

"One hundred thousand dollars in cash 'to you'."

"Looks like he's offering cash dividends to the public on his five syndicate wells in Santa Fe Springs. He's claiming he will pay a dividend on February 15th to all unit holders of record January 31st. Listen to this." "The wells on those syndicates have been in production nearly three years and up to date have paid approximately one million five hundred thousand dollars in cash dividends." Now listen to this, he claims thousands and thousands of people have already made over two hundred percent on their investment at these syndicates and most of the wells are still producing and should be for many years to come. Miss Carr, my son has ten units of Julian Petroleum. He's never received a penny. Look at this: "Units in all of these syndicates for months past have been selling on the open market for less than this one dividend on each unit will amount to,' which all goes to prove that one never knows does one?"

"No, one never does, unless one keeps posted, which by the way, is food for thought for Julian Petroleum Security is showing all the earmarks of being a 'whip'." And now he becomes a poet. "Any old cat can be the cat's whiskers but it takes a Tom Cat to be a cat's pa'."

< 67 >

"You know Miss Carr, I would be willing to wager Julian has a Tom Cat up his sleeve. He's trying to boost the market price of Julian shares for some reason. Did management give you the O.K. to take this ad? They had him black balled last year."

"Mr. Bond checked with the downtown dailies. They are running the ad so he told me to go ahead and take their money."

"Thanks, Miss Carr."

Hank Bond entered Bill Feeney's office.

"Bill,I just got a call from the resident agent of the Treasury Department. There will be a press conference at the U.S. Court House, something to do with Julian Petroleum. Get over there and cover it. It will start in about forty-five minutes."

Bill Feeney grabbed his hat and coat and ran downstairs. Raul Dominguez was in the dark room. Bill knocked at the door.

"Raul, let's go. The Treasury Department is calling a press conference. The boss wants us there."

"I'll be right out. I'll meet you at your car."

Feeney parked near Olvera Street. He and Dominguez ran to the Court House, joining a group of reporters in the United States Attorney's Office. Standing at a lectern between an American flag and a flag used by the Treasury Department were three men who obviously represented the government.

"Gentlemen, my name is David Cahill. I'm a special Assistant United States Attorney from Washington. To my right is Postal Inspector Madeira and to my left is special agent Montgomery from the Treasury Department."

Bill Feeney interrupted. "Postal Inspector, how do you spell your name?"

"M. a. d . e. i. r a."

"Thank you, sir."

David Cahill continued, "Gentlemen, of continuing concern to the Department of Justice has been the activities in Los Angeles of Chauncey C. Julian, former President of Julian Petroleum. After an audit of Julian Petroleum's books, and with the cooperation of Mr. Dougherty, the California Corporation Commissioner, the Treasury Department has placed a lien on all known assets of Chauncey C. Julian, not the corporation. It is our contention that Mr. Julian avoided payment of seven hundred and ninety-two thousand dollars in income taxes over and above the amount he has paid. Mr. Julian has agreed to pay the government two hundred fifty-four thousand, twenty four dollars as taxes due on his 1924 income. He has also filed notice of appeal here at the United States District Court relating to our claim for taxes due of over one half million dollars for the years 1922 and 1923. We are confident that the government's position will prevail in the courts."

"O.K. Raul, did you get all your shots?"

"Yes, Bill."

"Let's get out of here and find out what Julian has to say."

Raul and Bill rushed to Bill's car as the other reporters pressed the government agents for more information. Arriving at Seventh and Broadway, Bill and Raul entered the Loew's State Building, taking the elevator to the third floor. Bill knocked at Julian's office door.

"Come in."

Julian was seated at his desk. Jim Beebe was busy at a table near the window facing Seventh Street with stacks of stock certificates.

"Well C.C. , we just came from the U.S. Attorney's office. Do you want to comment on your tax problem?"

"No, Bill, we settled the matter in part; the rest of the matter will be in the courts for years. Listen, would you two fellahs like to get in on the ground floor of Western Lead? I'll be starting my campaign next week. Gentlemen, you are about to be given the opportunity to cash in on a world wide shortage of lead. Check with Mr. Beebe right there and you two may be my very first customers?"

"Raul?"

"I'll pass, Bill."

"Gentlemen, never forget," Julian raised his right hand heavenward. "I have a project on hand that from a money making standpoint should make the biggest oil project look like a punched out meal ticket. Save your dimes and dollars gentlemen because even one dollar in Western Lead Mining Company should make you a nice piece of change."

"Thanks C.C., we'll be watching the lines outside your doors just like the good old days. Goodbye for now, old boy."

CHAPTER 10

"C.C., how are you?"

Chauncey Julian stood at the entrance to the lobby of the Biltmore Hotel. S.C. Lewis lumbered over extending his pudgy hand. Julian shook it causing a grimace on Lewis' fleshy face.

"I see your shadow is here S.C." Julian pointed his finger in the direction of a smirking Jack Bennett. Julian otherwise dismissed Bennett's presence. "C.C., can we chat for a moment?"

"Here, now, I'm in a hurry S.C."

Lewis displayed a peevish grin. Bennett remained silent, his jowls arched in a familiar subconscious artificial smile.

"C.C., we have some problems."

Julian stood erect, proud and disdainful. The balding S.C. Lewis moved his head from side to side.

"C.C., you are creating problems for me."

"S.C., your problem is Julian Petroleum Corporation, not C.C. Julian. I'm out of it totally."

"C.C. calm down. Julian Petroleum is doing much better, you know that."

"Yes, I know that. In February the preferred advanced nineteen dollars and fifty cents a share to twenty nine dollars a share."

"C.C., you're not listening. You are creating a problem for me. This Western Lead promotion—your ads—they are disturbing my banker friends."

< 69 >

"Which friends, S.C.?"

"You want names? They're important people. They're helping me. Henry M. Robinson and Motley Flint. We're working together to shore up Julian Petroleum. There may be a new Governor and that means a new Corporation Commissioner and you are not helping matters. You told me last year you would not initiate any new promotions in California. You promised."

Julian raised his arm slowly, pointing his finger at Lewis.

"I told you I would not promote any new oil ventures in California. I didn't say anything about lead. You tend to your knitting and I'll tend to mine. Good day."

Julian walked toward the main Biltmore entrance. His Rolls Royce was waiting, Jim Beebe at the steering wheel.

"Let's get back to the office, Jim."

S.C. Lewis turned to Jack Bennett.

Julian will never learn to cooperate with the powers that be. He doesn't understand its a matter of survival. The bankers here can't stand him. He never understood politics. He could have really been an important man in Los Angeles, but he either didn't understand the power game or didn't like it. We'll probably never know."

"I like his choice of women, S.C. You should have seen that gal he was with the other evening at the..."

"Shut up, Jake, let's get back to business."

"Bill, this is Hank, can you come over to my office?"

"Sure, I'll be right over."

Bill Feeney hung up. Hank Bond was smoking his pipe. Bill seated himself near his editor's desk.

"Listen Bill, this piece came in today on the wire service from the New York Evening Post. We will run it in today's business section. Basically, it's a report by a fellow named Vogelstein, Chairman of the American Metal Company Limited. He is an expert on all kinds of metal. He's making a projection on the metal markets using statistical reports for the first ten months of 1925. He claims we are in a period of expansion founded on three important political and economic events. One, he lists as the inauguration of the Daw's Plan; second, the election of President Coolidge; and third, the British election and formation of a cabinet in which the business world has confidence. Bill, I'd like you to interview Henry M. Robinson at the Pacific Southwest Trust and Savings Bank on his views of the positive influence of the Daw's Plan. He was a member of the first Daw's Reformation Commission to rehabilitate post war Germany. Also, Vogelstein claims that the development of the storage battery for automobiles and radios and the increasing use of underground lead-covered cables, especially telephone, telegraph, and power transmission has kept lead mines working full time. If this is true, it's possible that this new promotion of C.C. Julian may have merit. Did you see that ad he placed in the paper?"

"Yes Hank, Miss Carr in advertising showed it to me."

< 70 >

"Bill, it read like his old stuff when he was touting Julian Petroleum. I checked with the downtown papers and they decided to run his lead mining ad so I didn't see any reason for us not to run it. I'd like you to do a story on the Julian mines. Find out where they are. Get some pictures. Don't blow it up out of proportion, but let's verify what he is up to."

"Fine Hank, I'll get right on it."

Feeney returned to his desk and placed a telephone call to Raul Dominguez.

"Raul, could you come downtown with me to Julian's office. Hank Bond wants me to ask him some questions about his Western Lead Mining Company. We can take the Red Car and have lunch at Cole's Restaurant over on Sixth Street in the Pacific Electric Terminal Building. It's just down the street from Julian's office and I like their turkey dip sandwich."

"Ok, Bill, shall I take a camera?"

"No, I'm going to try to get Julian to take us to his lead mine. We can shoot some pictures out there, wherever it is. Meet me out on Wilcox."

Bill Feeney and Raul Dominguez walked to Hollywood Boulevard and Highland Avenue. They entered a Red Car which arrived from the San Fernando Valley, en route to downtown Los Angeles. The two men walked south on Hill Street, past the Biltmore Hotel and Pershing Square, turned left on Seventh Street to Broadway then right to the Loew's State Theatre Building.

On the third floor, a line of investors stood outside Julian's office. Below on the second floor, the Julian Coffee Shop was filled to capacity. Jim Beebe sat at a desk outside Julian's office exchanging Western Lead Stock certificates for Julian Petroleum shares. As Bill Feeney and Raul Dominguez edged through the crowd, Chauncey Julian entered the corridor. The investors cheered and applauded, "C.C...C.C." Julian raised his arm.

"My friends, I predict a world shortage of lead which should drive prices of lead up to fourteen or fifteen cents a pound this year. Folks, Western Lead Mining Company property was discovered by a death defying desert rat."

Bill Feeney interrupted, "Death Valley Scotty, C.C.?"

Julian turned to Bill Feeney not flickering an eyelid.

"No, not Death Valley Scotty, but if you, my friends, could only see with your own eyes what you as shareholders of mine will share in...why you would hardly believe it possible that any man would let you in on this basis. But I am letting you in on it, because you are my friends. If I could only bring you to a realization of what you stand to share in here in this wonder mine alone—I believe that it would be one of the happiest days in your lifetime. I believe this great opportunity should stand out in your mind like an illustrious scintillating diamond in a handfull of pebbles. I believe that those of you who are able to judge between the true and the false...the worthless and the meritorious will see in the amazing progress of our company the handwriting on the wall. Time and tide wait for no man. So if you want to get in here with me you had better take action this very moment. And that is final."

The crowd cheered and applauded. Bill Feeney approached Julian.

< 71 >

"C.C., may I have a word with you?"

"Certainly, Bill, come into my office."

"C.C., you remember Raul Dominguez our photographer at the Citizen News?"

"Certainly, come in, Raul."

Julian seated himself at his desk. On the desk were small piles of Julian Petroleum shares which had apparently been traded for Western Lead Mining shares. Also nearby on the desk was a batch of Pacific Stock Exchange sell order forms.

"C.C., my editor wants me to take a look at your lead mine."

"Your editor wants you to take a look at my lead mine."

"That's right."

"I suppose you want to take a picture."

"That's right."

"When, Bill?"

"Where is the mine, C.C.?"

"In Death Valley. Look, I've got to keep the momentum on the sale of my mining shares going strong. I think visiting the mine is a good idea. How about the first week of May?"

"That's two months from now. My editor won't go for that. Tell me where the mine is. Raul and I will meet you there."

"Very well, since you insist. Do you want to form a caravan? Some of my friends want to see the mine so we might as well all go out to the desert together."

"Good enough. How about Saturday leaving about five a.m."

"All right, Bill, meet me in front of my home. It's about a seven hour trip. The mine is off the Las Vegas road past Beattie in Death Valley on the Beattie Road. I'll have some picnic baskets prepared."

"Fine, C.C., see you Saturday. By the way, will Mary Olive be going with us? If she is, I'll ask Kathryn and Billy to tag along."

"Not a bad idea. Basically, we have sort of patched things up as they say."

Southeast of the Julian Lead mine, hidden in the recess of a rock horst, stood Corporation Commissioner Dougherty, a geologist, two of his investigators, and Postal Inspector Madeira.

"What are they up to, Commissioner Dougherty?"

The Commissioner steadied his powerful binoculars on a rock outcropping. Further up the Beattie Road were a group of automobiles which formed the Julian caravan from Hollywood to his Western Lead Mines. Commissioner Dougherty and Inspector Madeira followed the caravan unobserved by Julian and his companions. Dougherty watched the activities of the Julian party which included several reporters and press photographers.

"They're posing for pictures, Manny. Julian has his arm on the shoulder of that gunslinger Tex Boren, his bodyguard, and the other arm around that newspaper reporter's son. His wife is standing to his left, dressed fit to kill. Smiling. I don't get it. His girlfriend, Winnie St. Cyr, is

< 72 >

standing behind him. The damn fool investors are smiling like Cheshire cats. Pete?"

"Yes sir, Commissioner."

"I want you to get court orders from a federal judge to allow us to take an assay of the so-called ore on that piece of crap. Will you help us on that, Madeira?"

"Of course, Commissioner."

"I'm going to put that charlatan out of business if it's the last thing I do as Commissioner. The Dougherty family has owned property in Santa Fe Springs since 1897 and that four-flusher compromised all our potential leases with his Julian Petroleum promotion. I'm sorry I couldn't put him out of business last time, but this time I think I've got him. Look at him waving his hands...Now they're setting up a picnic. Can you believe the gall of that guy? Chet, what was the name of the engineer you lined up to do the assay?"

"Ira B. Joralemon."

"Get him down here as fast as you can after we get the court order. I want to move forward quickly and get his stock off the Los Angeles Exchange pronto. Chet, also try to find out how he ever got his stock listed."

"Yes, sir."

"Let's get out of here before Julian spots us."

"What did you think of the Julian mine, Bill?"

Raul Dominguez stood at Bill's desk at the Citizen News. He tossed a dozen or so eight by ten glossies on the desk.

"There's Chauncey Julian with his arm around your son's shoulder. Look at the crowd, he's their hero. You know, I have no photographs of mine equipment. There wasn't any."

"Raul, he announced to the crowd that we were standing on top of the biggest silver/lead deposit on earth and that within days he would have high grade ore on the way to the smelter. We will print a picture of Julian and his friends. In the story we are simply going to state that Chauncey Julian, as verified by a federal mine claim, recorded his purchase of fourteen abandoned mines in Death Valley for an undisclosed sum and we will indicate that when we visited the mine location, no buildings were sighted nor were any mine shafts being drilled."

"Bill, Julian didn't seem depressed that you forced his hand to take us out there."

"Of course not, he got free publicity."

"Bill, come over to my office." Hank Bond opened the door to his office, Bill entered. Hank Bond held an Associated Press wire release.

"Bill, this just came in from Associated Press. It says Corporation Commissioner Dougherty has ordered Chauncey Julian to appear before a hearing to show cause why the Western Lead Mining stock should not be removed from trading on the Los Angeles Stock Exchange. Dougherty intends to confront Julian about his assay claim at Western Lead Mining Company. Can you tie that in with your visit to Western Lead Mining mine location?"

< 73 >

"Certainly, Hank, and thanks."

"Raul, place a call to Julian for me and get his reaction."

"Mr. Julian, this is Raul Dominguez from the Citizen News. We would like your reaction to the wire service report that Commissioner Dougherty will initiate a hearing as to the legitimacy of Western Lead Mining Company?"

"Certainly. You can quote me. I intend to sue that Irish son of a biscuit eater for a sum of three hundred and fifty thousand dollars for persecuting me out of personal spite."

"Thank you, Mr. Julian."

CHAPTER 11

"Ladies and gentlemen, your attention please. Corporation Commissioner Edwin Dougherty has a prepared statement for the press relating to the promotion of Western Lead Mining Company, Mr. Dougherty."

Edwin Matthew Dougherty, perspiring in the July heat of an overcrowded conference room in the State Capitol in Sacramento, approached a bank of microphones. He wore a seersucker suit, blue and white which was blotched with perspiration stains. He read from prepared notes.

"First, I want to read to you from an advertisement placed in the Los Angeles metropolitan newspapers by Chauncey Julian, president of Western Lead Mining Company. This purports to be a statement made by his mine superintendent. 'C.C., you can tell the world for me that I believe we are sitting on top of the richest silver/lead deposit on earth, that within thirty days from today, I will have carloads of high grade ore on its way to the smelter.' You notice that Julian does not himself make this claim, he puts the words in the mouth of his so-called superintendent. The facts are that with permission of a federal court judge, my investigators, along with a reputed mining engineer, Ira B. Joralemon, personally investigated the Julian claim sites in Death Valley off the Beatty Road. They discovered that there is no mine. The property consists of fourteen unpatented, undeveloped claims. Further, records we discovered disclosed that not more than seven thousand dollars had been spent on the property since Julian took over. Our experts found no ore assaying either at thirty dollars, fifty dollars, or ninety dollars a ton as Julian has claimed in his newspaper advertisements and radio broadcasts monitored on his own radio station. We discovered absolutely no grade or bodies ever opened up. I am therefore ordering the trading of Western Lead Mining stock suspended on the Los Angeles Stock Exchange and the stock will be removed from the trading list. Any questions? Yes?"

"Sir, Casey from the Sacramento Bee. There was a story in a Southern California newspaper quoting C.C. Julian to the effect that you are perpetuating a vendetta against him and that he intends to file a lawsuit against you. Any comments?"

"Yes, I understand Mr. Julian has filed a lawsuit against me and against a Los Angeles newspaper. The matter has not yet been adjudicated. I have every reason to believe that the lawsuit will be dismissed. As to the

< 74 >

issue of a vendetta, since being appointed Commissioner by Governor Friend Richardson, I have tried to perform my tasks as Corporation Commissioner to protect the investors of California. Mr. Julian's business practices since I took office in 1923 have not been, in my judgement, in the best interests of the investors. He has become rich and there is no evidence that he has shared the wealth with his stockholders. On the contrary, the stockholders frequently act like sheep with Julian as the shepherd. He has tried to make me look like a bad wolf. Someday, and I hope it is now, Chauncey Julian will get the message—Get out of California!"

Dougherty wiped his brow, sipped some water and was ushered from the podium by his attorney.

"The time is 6:00 p.m. and now, once more, from our studios at radio station KMTR in Hollywood we present the Julian Hour. Here with his message for today is Mr. C.C. Julian. Mr. Julian."

"Thanks, Jim Beebe, good evening ladies and gentlemen. You surely have heard about the action of Commissioner Dougherty concerning Western Lead Mining Company. Well folks, I'm here to tell you I have something to offer you. This is going to be the only real legitimate opportunity to make something worthwhile I have had to offer you in the last six years. My friends, my troubles are caused not by me or my ideas, but by conspiracies to ruin me. Commissioner Dougherty has persecuted me out of personal spite. The newspapers are out to get me and won't accept some of my advertising. Well, here I am talking to you on the free air of the radio without a physical conductor between me and you, the listener, with my radio station handling current by the kilowatt. It starts with a simple receiver, your radio and culminates with my powerful transmitter capable of spanning all of Southern California; all these features and many others make KMTR valuable to all of you. And now folks, bend your ears. Listen to this. I'm going to expand your Western Lead Mining stock. I have purchased a mine in Arizona containing gold, silver, and copper for one million dollars. I have organized Julian Merger Mines, Incorporated. I am offering to exchange Julian Merger Mines shares for your Western Lead Mining Shares and I'm offering one million two hundred eighty six shares of Merger Mines beginning at one dollar per share. My office will be open September sixteenth to begin trading and sales. Don't miss the action folks. This is your friend C.C. Julian bidding you goodnight, and now we'll continue with the music of the Hollywood Hillbillies."

The day following the broadcast, an aide to Commissioner Dougherty entered the Commissioner's office.

"Commissioner Dougherty, we are being flooded with phone calls from stockholders of Western Lead Mining Company. Julian was on the radio last night with a new stock offer called Merger Mines, Incorporated. He's offering to sell the shares of Merger Mines or trade them for the old Western Lead Mine shares."

"He has no permit to sell those shares. Send two of our investigators down to Los Angeles and arrest one of his salesmen. I am going to get the Julian operation into the courts."

< 75 >

"Commissioner, what will happen if C.C. Young wins the election in November?"

"It's simple. I'll be out and my successor will have to deal with Julian. Forget about that, let's get the crooks into court where they belong. They really should be in prison. The rumor is out that C.C. Young will appoint H.L. Carnahan as Commissioner. I'm not sure the Cryer administration will go along with that. They're putting all their money on C.C. Young and they will have a lot to say about who takes over this office."

On November 22, 1926, the day after the election of Governor C.C. Young, Bill Feeney exited the elevator on the third floor of the Loew's State Building. There were two lines of investors outside Julian's office. In one line were people holding Western Lead Mining stock certificates; the other line appeared to be cash purchasers. Seated at a table outside Julian's office was Jim Beebe, a notary public and several cashiers. Tex Boren, Julian's bodyguard, stood before the closed door of Julian's office.

"Good morning, Tex, I'm Bill Feeney frown the Hollywood Citizen News. We met out at the Western Lead Mine. Is C.C. available, I'd like to ask him for his comment on the election of Governor Young."

"Yes, I remember you. Just a second."

Tex Boren, reputedly a former Chicago cab driver, was balding and fat with an oval face and jowls resembling a bulldog. Tex entered Julian's office and closed the door. A moment later he reopened the door and motioned Bill Feeney to enter.

"Good morning, C.C." "Morning, Bill. How's the Mrs.?"

"Just fine, C.C., working a few bit parts. Billy is growing up. Listen C.C., for the record, your comments on the election of Governor Young."

"There's a good side and a bad side. The good side is Commissioner Dougherty is out. Do you know he had one of my salesmen arrested for selling stock without a permit? Some guy named Carnahan was supposed to get the job, but Young appointed a fellow name C.H. MacMillan. I never bumped into him. The bad side is the L.A. bankers and my not-so-illustrious successor at Julian Petroleum got their man in the Governor's chair. S.C. Lewis is all mixed up with bankers and politicians. He's lost control of Julian Petroleum."

"That brings up another question C.C., my editor would like some pictures of your gold mine in Arizona. Could you arrange for the press to meet with you at the mine?"

"Certainly, Bill, the sooner the better. This time you will see some real action. How about some time in March?"

"That's fine, C. C., give me a week's notice so I can arrange for transportation and a photographer. I've got to get back to the office. Thanks for your time C.C."

"Any time, Bill."

"Bill, get downtown." Hank Bond called to Bill Feeney from the door of the editor's office. "The wire service notified us of a press conference called by the new Corporation Commissioner, Clifford J. MacMillan. It will

< 76 >

be in the Athenium Room at the Biltmore an hour from now. His subject is the Julian Merger Mine Promotion."

"Right Hank, do you want pictures?"

"Yes, take Raul with you."

Bill Feeney, with Raul Dominguez in the front seat, drove to downtown Los Angeles, turning right off Hill Street down the ramp to the Biltmore garage. Bill Feeney and Raul Dominguez climbed the stairs to the Mezzanine meeting rooms entering the crowded Athenium Room. Bill Feeney noticed three individuals who seemed out of place among the reporters. One was Jack Bennett, C.L. Lewis' right hand man at Julian Petroleum Corporation, another was Kent K. Parrott, a behind-the-scenes engineer of the Mayor Cryer administration, and the third individual was the Chief Deputy and pal of District Attorney Asa Keyes, Harold "Buddy" Davis. Feeney joined the group of reporters while Raul moved to the front of the room to the left of the podium joining several other photographers.

The newly appointed Commissioner MacMillan stood at the podium. Tall, businesslike, stern, he motioned for silence.

"Upon succeeding Commissioner Edwin Dougherty, for whom I have the highest respect, I have noted that the former Commissioner gave top priority to the attempt to clean up the promotion field in Los Angeles. The good name of the city of Los Angeles is at stake. It continues to be necessary to publicly expose stock swindlers or the city will be stigmatized as a den of 'bunko artists.' Eastern newspapers refer to Los Angeles as the 'Paradise for Swindlers'. To a certain extent, the people who invest in these schemes are themselves responsible for being victimized. Were they to take a first thought or even a second thought they would not be taken in by falsehoods of oily salesmen and ballyhoo advertisements in bold type followed by ridiculous rhymes. When excessively great returns are offered, the investor should ask himself, 'Why does the man want to sell? Who is the man? What are his business qualifications, if any? What commercial standing has he in Los Angeles, if any?'

I am referring specifically to Chauncey C. Julian and his latest charade: the so-called Merger Mines. I personally have not been associated with the previous inquiries into Julian by state and federal authorities. However, we have obtained a court order forbidding Julian to sell Merger Mine stock. Further, I am today demanding that the Los Angeles District Attorney, Asa Keyes, criminally prosecute Chauncey Julian for his blatant unlawful sale of Merger Mine stock, and I strongly recommend to the Los Angeles newspapers that they cease taking ads for Merger Mines Incorporated, or his latest scheme, the new Monte Cristo Mining Company, for which I will issue no permit to sell stock to Chauncey Julian or anyone else."

As the Commissioner prepared to take questions from the reporters, Bill Feeney moved unobtrusively to the front of the room. He tapped Raul Dominguez on the shoulder, motioning him to follow him out the door of the meeting room into the large hallway of the Mezzanine meeting room section of the Biltmore.

"Raul, did you get your pictures?"

"Sure, what's the rush?"

< 77 >

"I want to find out what Jack Bennett is up to. Leave your camera in the art gallery and when Davis comes out of the meeting, follow him. Meet me back here at twelve thirty in the bar. I'm going to follow Jack Bennett and try to discover why he is associating with Davis and Parrott."

The two men walked down the steps from the Mezzanine area. Raul Dominguez placed his camera in the Art Gallery.

"I'll be back in an hour Gloria, watch it for me."

"Sure, Raul," responded the Gallery sales clerk.

Bill and Raul stood just inside the hall leading to the Biltmore Bowl.

"There they go."

Jack Bennett and Buddy Davis headed for the exit to the main lobby. Feeney followed the two men at a distance. Bennett and Davis walked past the bank of elevators, down the staircase and exited the hotel at it's Hill Street entrance. Bill watched them turn left crossing Hill Street to Pershing Square. Bennett entered the Pershing Square building on the corner of Fifth and Hill. Raul used the Fifth Street exit from the hotel and stood on the Northwest corner of Fifth and Hill. Buddy Davis stopped for a moment at the entrance of the Pershing Square building, said something to Jack Bennett and continued walking north on Hill Street. Feeney sat on a bench in Pershing Square, across the street from the Pershing Square building, the Julian Petroleum Corporation headquarters of C.S. Lewis. Raul walked west on Hill Street while Buddy Davis on the opposite side entered the Clark Hotel. Raul jaywalked through the heavy traffic and entered the Clark Hotel's lobby, Buddy Davis entered the coffee shop. Raul stood as though waiting for a shoe shine. Five minutes later Buddy Davis left the coffee shop. Raul followed Buddy north on Hill Street then east on Sixth Street to Spring where Davis entered a men's clothing store at 609 Spring Street. Raul stood in a doorway nearby after buying a newspaper. An automobile with a Los Angeles County logo on the door drove up to the curb. Raul watched the District Attorney, Asa Keyes, enter the clothing store. Raul slowly walked by the front window; Buddy Davis and his boss Asa Keyes were not visible. Seated in front of a curtained partition toward the rear of the clothing store was a tailor working at a sewing machine. Raul checked his watch. It was 12:10 p.m. Raul tossed the newspaper into a rubbish bin and headed back to the Biltmore Hotel. Feeney motioned to Raul from Pershing Square. "Raul, look over there." Bennett and Lewis were walking east on Fifth Street. Bill and Raul walked on the Pershing Square side of Fifth, keeping a distance of thirty paces behind Lewis and Bennett who then proceeded to Spring Street and entered the main office of the First National Bank.

"Raul, stand in line to make change and then come out and tell me what you see."

Several minutes later, Raul exited the bank and reported to Feeney.

"Listen, Bennett and Lewis walked to the sitting room outside the Bank President's office. Motley Flint was waiting for them. He chatted for a moment and then Henry Robinson ushered them into his office and that was that."

"Where did Buddy Davis go, Raul?"

"He met his boss, the D.A., Asa Keyes and they went to the back

< 78 >

room of a Haberdashery further up Spring Street."

"Something is going on Raul. Let's get back to the office."

Bill Feeney and Raul Dominguez entered the reception area of the Citizen News. They were greeted by the ever-smiling Maggie McCloud.

"Any messages, Maggie?"

"Yes, Mr. Feeney, Chauncey Julian called you from KMTR Radio. He said you have his phone number."

"Thanks, Maggie."

Bill Feeney climbed the stairs to the newsroom, walked to his desk and slouched in his chair, tired after the surveillance in downtown Los Angeles. Bill Feeney placed a phone call to Chauncey Julian.

"Hello, Hi Jim, is C.C. there?"

"Yes, Bill, I'll call him. He's working on a speech."

"Bill, this is C.C. I want you to do a story on this. My listeners are complaining that someone or something is interfering with my speeches on KMTR. They say the music is perfect but when I start my Julian Hour there is static and interference. I want you to know and I want the public to know that I am appealing to the new Federal Radio Commission in Washington D.C. I want a complete investigation. Someone is trying to silence me."

"Sure seems like a story, C.C. Ask some of your listeners to call me and fill me in on the details. By the way, how about the request by the new Corporation Commissioner to have you criminally prosecuted by the Los Angeles District Attorney?"

"If the Governor doesn't get rid of MacMillan in three weeks...I let the word out...if the governor doesn't get rid of him...I'll blow the lid off of what Lewis and his banker friends are up to with their pools of Julian shares. You don't need to print that, they all know how I feel."

CHAPTER 12

"Judge, I think most of the players are lined up. It's time to move quickly."

Jack Bennett sipped a cup of coffee, S.C. Lewis wiped his mouth with a white linen napkin as he rolled away a room service luncheon.

"Jake, I agree. That's why I moved my personal office from the Pershing Square Building to the Pacific Southwest Bank Building. This will distance me from you and Ed while you do volume preparation of the stock certificates. Also, it will separate you folks from the Corporation Treasurer Campbell and Secretary Conroy. I don't want them nosing into what you're doing."

"Judge, there's a growing possibility that we'll get caught. Rumors are already spreading around the stock exchange that there is an over issue of Julian stock."

"Jake, don't worry. Worry makes you look tired. Kent Parrot is calling in his markers on this. And there's a petition circulating requesting that Governor Young replace Corporation Commissioner MacMillan with the Los Angeles City Attorney Jack Friedlander. This will solve two

< 79 >

problems. One, when this bubble blows, he is personally acquainted with most of the players, at least the bankers and the movie people; and two, hopefully by getting rid of MacMillan we can shut up Julian with his damnable radio speeches. He promised me he would get out of promotions in Los Angeles. You can't trust his word."

"Come on, Judge, I heard him say he would get out of oil. He never mentioned mines."

"Never mind, Jake, you stick to your end of things. I tell you Julian is a problem. His little-guy investors are out there with legitimate shares and they're very loyal to Julian. Besides, Julian knows all about the stock pools. He gives them names. The Motley Flint pool, the Jewish pool, the Tiajuana pool. His investors are complaining to Julian. They say they are not in on the profits these pools are making from earnings and stock commissions. There is another bright side, Kent Parrot has a plan in the works to replace Friedlander as City Attorney with E.J. Lickley. He is a former Deputy D.A. and is Executive Secretary of the Church Federation. If things go sour, hopefully he will not embarrass his friends. The bankers want us to call a meeting of the stockholders at the Olympic Auditorium on Friday the twenty seventh so prepare a notice, but don't put my name on it. There will be an announcement of the merger of Julian Petroleum into the new California Eastern Oil Company. One share of Julian is to be exchanged for three and one half shares of the new company at ten dollars per share. The bankers have an audit to prove to the stockholders the dramatic growth in Julian Petroleum assets. On the same day before the meeting, if everything goes according to Parrot's plan, the Governor will name Jack Friedlander as the Corporation Commissioner. By the way, Jake, how are we standing cash wise?"

"Well, its not all ours, but its approaching twenty one million, some in numbered accounts. Well, Judge, I think its time to get back to work."

Bennett returned to his office at the Pershing Square Building. Lewis walked to his headquarters at the Pacific Southwest Bank Building on Spring Street. Entering the reception area of his office, he asked:

"Any messages worthwhile, Doris?"

"Yes, Mr. Lewis, you have an urgent call from John E. Barber of First Securities Company."

"Thanks, Doris."

Lewis entered his office, closed the door, and placed a phone call.

"John, this is Judge Lewis. How are you sir?"

Barber responded coldly, "Mr. Lewis, before we proceed with the financing of the purchase of Marine Oil Company, our board wants to verify one more time that they have a valid contract to supply oil to the Southern Pacific Railroad. Go ahead with your meeting at the Olympic Auditorium, but I want to caution you, we hear continuing rumors of an over issuing of Julian stock. Before we proceed to finance the merger to form California Eastern we want to audit your stock books."

"John, I can't do that. It would take too much time. Besides, the stock books are at our office in New York."

"I know they're in New York. I made arrangements to ship the stock

< 80 >

books with the help of the corporation stock transfer clerk. If you want to proceed with the merger, we must have those books brought here intact from New York. The books will remain in the custody of the stock transfer clerk and our security people. They will remain locked up until Mr. Kotteman, our auditor, completes his investigation. Also, as you well know, if this merger is successful, Harry J. Bauer will act as Chairman of the Board of California Eastern. He has personally invested over one half million dollars of his own money to help steady the potential merger of Julian into California Eastern."

"Very well, Mr. Barber, but remember I am still President of Julian Petroleum Corporation and I will release the stock books to your auditor on my terms."

"Listen, Lewis, don't threaten me. The stock books will remain with the transfer agent until Kotteman completes the audit. Our people are already in your New York office. Have your stock transfer clerk proceed immediately to transfer the stock books. Good day, Mr. Lewis."

Lewis hung up the phone for a moment and then placed a call to his New York office.

"Hello, Harry, Judge Lewis. We have a little problem. I'm going to have to let the auditor at First Securities go over the stock books. Pat Shipps will be arriving in New York shortly. Start packing the stock books in trunks. I'll deal with the situation when the trunks arrive in California."

"Yes, sir."

Lewis then placed a call to Jack Bennett.

"Jake, its me. Step on it. Get as much stock out as you can. I'll stall the auditor as long as possible. I agreed to let them look at the stock books. They'll be in the custody of the stock transfer clerk and their security people. I'll see you at the hotel tonight. I'm going get ready for the stockholder's meeting."

"Mr. Barber, William Kotteman to see you."

Responding to his secretary's message on the intercom, Barber rose from his chair greeting the bank's auditor with a handshake. Returning and seating himself at his desk, John Barber lit his pipe as Bill Kotteman took some papers from his valise.

"Mr. Barber, the situation at Julian Petroleum is a catastrophe. First, let me tell you that it was like pulling teeth to get S.C. Lewis to allow us to audit the stock books. There are eleven trunks full. We kept them locked up in the transfer agent's garage. He accompanied the trunks along with our security guards. Mr. Barber, there has been an enormous over issue of stock. Before I give you the numbers, I want to interject something that Shipps told me. Apparently, in March of 1925, Jack Bennett ordered him to issue a new stock certificate for ten or twenty shares of Julian without providing an old certificate. Bennett told the transfer agent that he would deliver the old certificate for cancellation the following week. This situation prevailed, with Shipps keeping tabs in a separate journal he called 'the Jack Bennett shares'. He lost count, but his best guess is that about a million 'Jack Bennett shares' had been issued through April of 1926. Anyway, as a result of our audit up to this date of May 6th, the authorized number of preferred shares is six

< 81 >

hundred thousand, however there are three million six hundred and fifteen thousand shares of preferred outstanding. Of the common stock, there are one million, two hundred and seventy eight thousand more shares outstanding than were authorized. Based on an average market price of ten dollars a share for the preferred, receipts from the preferred and common appear to be near thirty three million dollars in two years. We also know that the corporation has eleven million in bank loans outstanding. Lastly, their profit from their brokerage house, the one they purchased, Wagy & Co., approached two million dollars last year. A previous audit of the corporation assets shows slightly over sixteen and one half million dollars net worth. There appears to be the strong possibility of a substantial amount of money unaccounted for."

Barber turned in his chair staring out the office window as the first shades of evening silhouetted the office buildings on Spring Street. Regaining his composure, he reached for the office intercom and pressing a lever he calmly requested his secretary to place a call to C.S. Lewis.

A few moments later Lewis was on the line.

"Mr. Lewis, this is John Barber. We have a serious problem. We need to talk immediately."

"Very well, your office or mine?"

"Mine, our auditor is here with me."

"I'll be over in five minutes."

"Fine, I'm going to notify the press that you will be in your office to make a statement at 7:00 p.m. That will give us time to chat before the reporters arrive at your office."

"Bill Feeney, here."

"Bill, this is Hank. Listen, we just got a telephone call from John E. Barber's secretary at the First Securities Company. She indicated that S.C. Lewis is calling a meeting for the press at his office at 7:00 p.m. Get downtown as fast as you can. It's 5:30, can you make it?"

"It'll be close, I'll leave right away."

Feeney jumped up from the easy chair in his small office where he had been reading the afternoon paper. He called to his wife.

"Kathryn, I have to go downtown...in a hurry...no time for dinner...love you."

Bill Feeney jogged to his car, taking the Riverside Drive route to downtown Los Angeles rather than the Cahuenga Pass through Hollywood. He parked his car near the Southwest Bank building and rushed to the third floor office of C.S. Lewis. A cadre of seasoned reporters from local dailies plus eastern newspaper stringers were already gathered in the reception area. The strong odor of scotch permeated the room due to the late hour of the conference. A few photographers were present. A buzz from the speaker on the receptionist's desk signaled her to open the door to the corporation president's office. Lewis appeared in the doorway alone. His countenance was composed, but anxious.

"Gentlemen, under advise of our corporate counsel and with the concurrence of officials at First Securities who are handling financing for our

< 82 >

anticipated merger with Marine Oil Company into California Eastern, I have this evening formally requested that the Board of Governors of the Los Angeles Stock Exchange suspend trading the stock of Julian Petroleum Corporation. An audit must be completed to determine the extent of an apparent over issue of stock. I wish to add that legal action will be taken to restrain transfer of the stock and that, to the best of our ability, we will take financial and legal action to protect the interest of every stock holder until the exact situation relative to the over-issue is determined. This move will not deprive recent stock purchasers of any rights, but it will effectively close the books of the corporation to enable the auditors to complete their audit. I have also informed the office of Corporation Commissioner Jack M. Friedlander of our decision. I have directed our corporate counsel to seek in federal court a temporary injunction restraining all transfer or retransfer of Julian Petroleum Corporation Stock as a means of protecting the stockholders. I'm sorry gentlemen, I cannot respond to any questions. I suggest you contact Mr. Friedlander tomorrow."

Lewis abruptly closed the door to his office. The reporters were stunned and scurried out of the reception area toward the elevators and pay phones to file their stories. Bill Feeney placed a call.

"KMTR Radio, Jim Beebe speaking."

"Jim, this is Bill Feeney. Is C.C. in?"

"Yes Bill, one moment, he's in the studio."

Beebe activated the red light in the studio. Julian picked up the phone.

"Yes, this is C.C. Julian."

"C.C., this is Bill Feeney. S.C. Lewis just announced that he has asked the Los Angeles Stock Exchange to suspend the trading of Julian Petroleum Stock. Any comment?"

"Not about the suspension, I knew it was coming. As soon as I get the facts put together, I'll do a special broadcast on the Julian Hour. You can print that. The date will be, let's see...May fifteenth at 6:00 p.m. Someone keeps trying to block the airways during my broadcast. I tested the transmitter and its functioning perfectly. You can add to your story that to reach my people I intend to install loud speakers on Seventh and Broadway in downtown Los Angeles so the folks can get the truth about this mess. I intend to tear the veils from the affairs of Julian Petroleum Corporation. The story I will tell will be red hot. I am going to name names."

"Got it, C.C., I'll add your statement as a follow up to the suspension story in tomorrow's paper."

CHAPTER 13

Julian and Boren, with Jim Beebe as chauffeur, drove from the mansion in Hollywood to his office at the Loew's State Building. At the third floor office, Julian was confronted by a crowd of frightened, disheartened stockholders—his folks.

"C.C., what am I to do. I've lost everything," an elderly woman

< 83 >

cried out.

"There, there my dear. Next week on my radio broadcast I'll place the blame for the destruction of Julian Petroleum where it belongs. I'm going to name names."

"C.C., there are thousands of people over on Spring Street outside the stock exchange. They don't know what's happening."

"Folks, on the evening of the fifteenth at 6:00 p.m., I will place loudspeakers on the outside of this building. I'm asking you folks to come down here and I'll identify the people who caused your ruin. Someone is interfering with my radio programs so I'll speak both over the radio and from the loudspeakers. You will find out the whole truth about Julian Petroleum."

"Wait a minute, C.C.. Keep quiet everybody, and listen for a minute!"

A tall man dressed in a dark suit pushed his way through the crowd directly in front of Julian. Boren lunged forward in front of Julian, but Julian motioned him to stand back.

"Speak as you wish, sir."

"Listen, C.C., are you going to tell me that you're going to be the Saviour of Julian Pete? The first time I bumped into you it cost me four hundred dollars for some adventure you were having. You wanted to bore a hole in the ground and put Union Oil out of business. You supplied all of us with the finest line of glittery verbal prune juice to be found in any dictionary. You tapped the savings of all of these people here and all of the widows and orphans on the block. Whenever you went broke you started some new merger. When your oil went rancid you dreamt up a lead mine and people like me poked money under the door of that office to get a piece of it. You know you can't peddle your wares anywhere in the world except in Los Angeles. We're all a bunch of softies and you know it. Why don't you go down on Spring Street right now and try to sell the crowd the Pacific Ocean for a dollar a bottle as eyewash for their tears. You sent me an eighteen page letter not long ago saying you needed a million dollars for your Monte Cristo Mine. You named your so-called mine after Dumas' hero, because at one time he thought he had all the money in the world. Well, here is a check for one million dollars. About ten dollars of it is good. The rest will be all right if you can get the endorsement of William Rigley. That should be easy. Tell him I'm one of your little guys."

"That's enough," grunted Tex Boren who pushed through the crowd to Julian's office. Jim Beebe and Julian followed. Julian chatted with several of the familiar faces among the crowd of victims of the Julian Pete crash.

Seated at his desk, Chauncey Julian turned to Jim Beebe.

"Jim, call our attorney. Tell him to review the liable law. Tell him I'm going to name a lot of names on the radio on the fifteenth; big guys, bankers, respectables, people responsible for wrecking Julian Petroleum. Have him give us a run down on liable as it applies to radio broadcasting. Tex, let me know when the crowd thins down outside and we'll go over to KMTR. I want to start working on my speech. It's possible that if those disgruntled people start having meetings, things could get rough. I want to be in a position to announce a desire to form an association of Julian Petroleum Corporation stockholders to help protect the innocent investors. I want to be like Moses leading my folks, the little guys, out of the mess that the bankers created."

"Mr. Julian, the attorney is on the line. He wants to talk to you about that question concerning naming names on the radio."

"Mrs. Julian, there are two gentlemen on the front porch asking to see you. One says he is your father and the other claims to be your brother."

Mary Olive Julian sat in her favorite chair in the library. The maid, Margaret, seemed surprised by the visit, but not perplexed as she knew nothing of her employer's family background.

"Margaret, my father is deceased and I have no brother. What do these men look like?"

"Well, Mrs. Julian, both men are very tall. They're wearing cowboy boots. Each is carrying a Stetson hat in his hand. They have Texas accents."

"Tell them to come in, Margaret."

The two men inched their way into the library. Mary Olive Julian stood near the library window.

"Good afternoon, gentlemen. I am Mrs. Julian."

The two men stared at one another quizzically. The elder of the two responded, "Good afternoon, Ma'am, I am Jack Smith and this here is my son. We was told that my daughter June Smith was married to Mr. Chauncey Julian. We came from Texas for a visit, Ma'am."

"Well, I'm sorry gentlemen, I am Mrs. Chauncey Julian and have been so for many years. Gentlemen, my husband travels a great deal with your daughter and your sister, sir. I suggest you contact Mr. Julian at his office where you may telephone June. Here is her phone number."

Mary Olive Julian walked over to her desk and from memory jotted down a phone number on a slip of paper handing it to June's father.

"Good day, gentlemen."

"Good day, Ma'am," the two Texans responded in unison as they backed out of the library.

The maid escorted the men to the front door where they hastily retreated toward their small Ford truck.

For one week Chauncey Julian played recorded announcements between sets of cowboy songs, the usual program fare at KMTR.

"Listen to your radio, folks. Sunday, May fifteenth, at 6:00 p.m. I'm going to tell the inside story of the ruin of Julian Petroleum Corporation. I will spare no one. Some of the biggest financiers in Los Angeles will be named and labeled. Or, come on down to Seventh and Broadway. I've installed great big voice magnifying horns. You won't miss a syllable of what I say. Bring your notebooks. Write the names down. They are the phantom power brokers of Los Angeles who think they can control this city and everyone in it, especially us little guys. Remember, 6:00 p.m. sharp. KMTR Radio or downtown, Seventh and Broadway. This is C.C. Julian, goodbye for now."

< 85 >

"Jim, you go on downtown and take care of the voice magnifying horns. Test them out and let me know if there are any problems. Tex, get up on the roof and make sure no one tampers with the transmitting equipment. Take your .38. Let no one on the roof. Tell our announcer to keep up my recorded messages between every three records all day. Do you understand? Also, when I go on the air tell the announcer to sit outside the studio door and admit no one. Got it."

"Yes, sir."

"C.C., do you want me to shoot the first bird that walks on the roof of this here building?"

"No, Tex, but if anyone shows up, frighten him and find out who paid him."

At five 5:55 p.m., the KMTR announcer played the last recording of the evening, "Who left the lock off the hen house door." The announcer read the latest weather report and precisely at 6:00 p.m. pointed his index finger at Chauncey Julian.

"Good evening, folks, this is C.C. Julian. When I left the president's job at Julian Petroleum Corporation, those wells were still gushing. We had our refinery. We had our filling stations. Most of you folks owned a real part of the company. You little folks. I let you in. My friends, you all have the realization of what it meant to be a shareholder in the Julian Petroleum Corporation. My friends, I know you had that realization. I remember fondly when I stood out there in Santa Fe Springs on the derrick floor and saw the monstrous, roaring, bellowing gusher burst in and roar out it's fortune song. My heart swelled within me with happiness and pride! Ah! What a thrill it was to see the raging green fluid burst forth with another fortune in the making. But it was not alone for myself that I thrilled with pride and happiness...it was also for you, you who had faith and confidence in me. Friends, I have emblazoned my name among the petroleum pioneers. Now I want my friends to know the facts, the facts about the ruin of Julian Petroleum Corporation. First, how did they ruin Julian Petroleum? They bled it white through loans at usurious rates of interest, through bonuses, commissions, and extorted gifts. They used their pools of phoney stock to play with while you, my folks, who bought your shares from me, were left out in the cold. Why did they do this? It's because they play a phantom game of politics, back door bribery and usury to pad their own pockets and further their political schemes. Who are they? I'll tell you who they are. They are the bankers of Los Angeles who would not lift a finger to help me when I ran the company. I wasn't part of their group. They didn't want me in their lodge halls or private clubs or their country club locker rooms. They played a game with all the little guys. They look at you and they call you 'the tax payers.' Well, we're all really the chumps for letting them get away with their closed door schemes.

They are human parasites, unscrupulous blackhearted widows. They are ulcers of the community.

Now, I'll name some names. These are the people who brought ruin to Julian Petroleum Corporation. First there's Charles F. Stern, President of Pacific Southwest Trust and Savings Bank. He is your former State Banking

< 86 >

Commissioner. Next there is Motley H. Flint, the brother of your former U.S. Senator and the Attorney for Julian Petroleum Corporation, Frank Putnam Flint. Another one of your looters was your ultra-conservative, super-patriot big shot in the anti-red Better America Federation, Harry M. Haldeman. And lastly, there is that great protector of people's investments, John E. Barber, president of First Securities Properties. There are more. I will name them next Sunday evening, same time, same station. Until then, this is your friend C.C. Julian. Good evening, everybody."

The crowd of about fifteen thousand people began to disburse, murmuring among themselves. Traffic was snarled for blocks around Seventh and Broadway. Trolley cars stood in lines.

The following morning when Julian returned to his office at the Loew's State Building there was a curious silence. No crowd. No newspaper reporters. No phone calls. The following Sunday Julian repeated his allegations naming several more bankers including the President of First National Bank. Again, Monday morning there was silence. But at 9:30 a.m., a call came from the City Prosecutor.

"Mr. Julian, this is Dr. Lickley, the City Prosecutor. We had a court reporter transcribe your speech on May fifteenth and also the speech yesterday. I am requesting that you come to my office today at 2:00 p.m. to substantiate the statements you made about citizens of Los Angeles. I want you to prove your statements and rename those whom you have charged."

"Sir, I will be glad to do so and also I trust that the Grand Jury will call me so that I may testify in front of them."

"That issue is in the purview of the District Attorney, Asa Keyes. He and his Chief Deputy, Davis, are out of town. However, I am certain that the Grand Jury will call you at the appropriate time."

"Gentlemen of the press, we appreciate your presence here in the board room of the First National Bank of Los Angeles. I am Harry M. Robinson, President of the First National Bank of Los Angeles and Chairman of the Board of Pacific Southwest Trust & Savings Bank. I have a prepared statement dated this date, June 27, 1929, Los Angeles, California. The statement is as follows. I have been home ten days. This time has been employed in an endeavor to develop the facts of the relationship of the Pacific Southwest Trust & Savings Bank to Julian Petroleum Corporation affairs.

"I have not been interested in any way directly or indirectly in the pools, outside loans, or bonuses and have not at any time directly or indirectly bought, sold or had any interest in the stock of Julian Petroleum Corporation, or any of the other corporations involved.

"In part, at the request of one of our important customers, we began the financing for the combining of certain oil properties, including the properties owned by the Julian Petroleum Corporation. These properties were valuable and of greater value when combined than when standing alone.

"Could the project have been carried through, it would have been of benefit to the stockholders of Julian Petroleum Corporation. This undertaking, in a sense, parallels the efforts of some of our people in connection

< 87 >

with the Los Angeles Investment Company where, when its difficulties climaxed, they took hold and built out of it a wholesome and prosperous institution. This was the intention of the people in our banks in this instance.

"The only way that I can see now for the stockholders of Julian Petroleum Corporation to realize anything would be through the final carrying through of a combination and refinancing substantially on the lines of the plan contemplated by the First Securities Company.

"We took as securities for the loans advanced the properties which we then and now believe was adequate for security purposes.

"The First Securities on its part undertook to build a new financing plan for the merger of the properties and were able to obtain partners in the underwriting only with the greatest difficulties because of the unusual reputation of Julian Petroleum Corporation.

"From the beginning, the attitude of our people has been constructive and had it been possible to have completed the plan, it would have been the ultimate great benefit to the community and Julian stockholders. We have been accused of ulterior motives and of the desire to make a profit out of the situation. This is absolutely false. All of the transactions of our two banks in this whole matter involving loans has been at the going rate of interest not exceeding seven percent and without bonus. In this connection, no loans were made by the banks, or First Securities Company, through any subsidiaries or other agencies.

"If the participation of any of the officers of the Pacific Southwest Bank in so-called pools is in question, it is their private affair involving their own money and has nothing to do with the bank's resources.

"It is perfectly clear that the whole trouble grows out of the over issue of the Julian Petroleum stock.

"None of our people had any knowledge of such over issue until the final audit disclosed the fact.

"We were then forced to tell our partners in the underwriting and the public of the fact of the over issue. I do not know how the public for a moment can fail to see that the cause of the trouble, which is perfectly obvious, is the over issue of stock and nothing else.

"It is equally obvious that we would not have gone forward in making bank loans nor in an attempt to finance the California Eastern Oil Company for the purpose of combining the properties had we thought for a moment that there was any over issue of stock.

"In addition, on behalf of the directors, we regret exceedingly that certain officers of the Pacific Southwest Trust & Savings Bank have been indicted by the Grand Jury. These officers have placed their resignations in the hands of the board to take effect at its pleasure.

"These officers have been with this bank for many years and have always borne unblemished reputations and have had our complete confidence.

"It is a cardinal principle of Americanism that everyone is considered equal before the law and that no one by virtue of wealth, position or condition is entitled to any privilege or protection denied to his fellows. If these men are guilty, we will join with the community and demand that they

< 88 >

be adequately punished. If they are innocent, they should promptly be exonerated. They and all other similarly accused are entitled to ask that the public suspend judgment as we do pending the findings of the court. Thank you, I have nothing else to say."

C.C. Julian sat at his desk reading the statement of Harry A. Robinson printed in the form of a half page ad in the morning paper. The phone rang.
"This is C.C. Julian."
"Good morning, C.C., this is Bill Feeney. Any comment on the statement of Mr. Robinson?"
"None whatsoever. I gave my statement to the Grand Jury. I talked to the City Prosecutor. I'm leaving for Cuba and Mexico before the fourth of July with a young woman, possibly the future Mrs. Julian. Tell your wife to be careful of Mary Olive. She has some kind of a mental problem. I'm going to have her treated when I return. Take care Bill and remember what Death Valley Scotty told us, 'Play to the crowds'. I'll see you later."
"Just a minute, C.C. A usually reliable source told Raul Dominguez that prior to your first radio broadcast, you sent Jim Beebe to speak to friends of the bankers. They claim Beebe informed them of the nature of the charges you were about to make. He allegedly told them you would cooperate with them, be a good dog, and say nothing if they would use their influence to put you in a position to sell your new mining stock legally and open the Los Angeles newspapers to your advertisements. Any comment?"
"Sure, Jim Beebe is no longer my employee. I have no idea what he may or may not have said. I will not get in a pissing contest with skunks. See ya, Bill."

CHAPTER 14

Christmas Eve 1930, Oklahoma City, Oklahoma. Chauncey Julian escorted Barbie Bratton, his traveling companion, into Rothchilds Department Store. Pierce Arrows and Reos awkwardly lined the streets outside. Over a hundred Osage Indians; men dressed in blue pin striped suits, Pendleton shirts, shoulders draped with Indian blankets, women in long silk dresses, again shoulders draped with Indian blankets, were spending their head rights, all minor oil barons. Julian had returned from his two year sojourn in Cuba and 'Mexico. He retained his apartment in Oklahoma City, and had opened an office for his new enterprise, C.C. Julian Oil & Royalties Company, in the Cotton Exchange Building.
In the women's fur coat section Julian watched as Barbie Bratton adjusted a full length Hudson Bay Company sable coat to her slim body. Two days previously Julian paid cash for the coat which required minor alterations plus embroidering the initials B.B. on the pocket of the silk liner.
"Do you like it, my dear?"
"Oh yes, Chauncey, I love it."
She kissed Julian on his forehead. Satisfied with the fit, they walked

< 89 >

outdoors. Julian hailed a taxi.

"Driver, take us to the Del Mar Gardens." Julian held Barbie's hand.

"Barbie, after New Year's, we'll return to Los Angeles. I'm taking possession of an airplane from Lockheed Aircraft in Burbank. We'll spend at least two weeks in Hollywood. I have to hire a pilot and cover a few loose ends. We'll stay at the Hollywood Roosevelt Hotel."

"Chauncey, what will my friends think? I told them we were going to be married."

"Now, now my dear. When I lost that court case to have Mary Olive declared insane, things came to a standstill."

"Well, you could have divorced her!"

"Not at all. If I had, she would have cleaned me out. I have to start a whole new operation here in Oklahoma. My sales people will be coming here from California. After the first of the year I start my mail campaign. Just be patient."

"Well, I don't like it here. I just may stay in California. I'm young. Look at me. I'm not going to grow old unmarried."

"Calm down, Barbie. We'll talk things over when we get to Hollywood."

The taxi pulled up in front of the Del Mar Gardens. Julian gave the cab driver a fifty dollar bill. Julian and Barbie entered the Del Mar Gardens just in time for the movie.

During the first two weeks of the new year 1931, Julian prepared letters and financial statements concerning the C.C. Julian Oil & Royalties Company. He used, in part, a mailing list of former investors and business associates in California. With that completed, Julian, Barbie Bratton, Tex Boren, and Julian's valet Joe Caddy arrived in California, by train, at the East Los Angeles station, to avoid possible reporters or disgruntled Julian Petroleum stockholders. Julian arranged for a limousine to transport the foursome to the Hollywood Roosevelt Hotel.

The following day, Julian and Tex Boren took a taxi to the new Burbank Airport where the Lockheed Aircraft manufacturing plant had moved from its old location in Santa Barbara.

Bankers in Oklahoma arranged a lease purchase agreeent with Lockheed Aircraft for one of the first Lockheed Orion single engine passenger planes constructed at the Burbank assembly line. For several days Julian sought the services of a pilot willing to move to Oklahoma. Tex Boren luckily located an old U.S. Army Air Corps buddy who was flying tourists to Mexico. He agreed to accept employment with the C.C. Julian Oil & Royalties Company. After finalizing paperwork, Julian, Tex Boren, and Arthur Riordon, the pilot, boarded the new plane along with a test pilot provided by Lockheed Aircraft Company. With the test pilot at the controls, the group flew in a circle over the San Fernando Valley gaining altitude, then over the Santa Monica Mountains north toward Santa Cruz Island. The pilot circled Santa Cruz Island dropping altitude above Santa Rosa Island, then flew landward arriving at the Santa Barbara Airport. A limousine met the aircraft and the four men, after the pilot arranged for refueling, were driven to the Santa Barbara Biltmore Hotel where Julian had reserved two cottages

< 90 >

for the weekend. At the registration desk, the clerk handed Julian the key to his cottage along with a telegram.

"Mr. Julian, this telegram just arrived for you." Julian opened the yellow envelope.

"Will not be returning to Oklahoma with you. My good name means more to me than your riches. Goodbye. B. Bratton."

Julian crumpled the telegram tossing it into a cuspidor.

"Tex, I'll meet you in the dining room in half an hour. Show the pilots their rooms in your bungalow."

Monday morning, the group flew directly fron Santa Barbara to Rurbank. Julian and Tex Boren arrived by taxi at the Hollywood Roosevelt Hotel at ten am.

"Are there any messages?"

Julian retrieved his room key and a folded note which he opened.

> From the desk of William Warren Feeney.
> To: C.C JUlian From: Bill Feeney.
> RE: Your Wife. Mary Olive has sued you for
> divorce. Claims you are worth two mil. Wants
> five K per Mo. Alimony. Read all about it in
> tomorrow's paper. Front page.

Julian stuffed the note in his pocket. He yelled at Tex Boren.

"Tex. Get packed. Have Joe Caddy pack my clothes in a hurry. Order a cab. Call the pilot. We're flying out of here at noon for Oklahoma. Don't ask any questions."

Bill Feeney parked his new Dodge in the driveway of his home in North Hollywood. The neighboring lots in the Toluca Lake area had sold rapidly over the years and his home was no longer isolated from neighbors. Bill picked up his copy of the Citizen News on the sidewalk near his driveway. He unlocked the front door and was greeted by his son, Billy, and Billy's dog, Gypsy, along with the part time maid, Mrs. Shafer, who cared for Billy when he returned from the Rio Vista School. Billy's mother had signed a contract with Universal Studios nearby, and now worked regularly in supporting roles.

Bill poured a drink, turned on the radio and began reading the paper. Bill's only contribution was a lackluster article dealing with a program to relieve unemployment in the City contrived by the Board of Water and Power Commissioners. They had hired a small army of two hundred and twenty five idle men who were put to work with picks and shovels digging a trench for a fifty four inch water line across the corner of the San Fernando Valley. During his second highball the latch on the front door clicked and Kathryn entered wearing a costume reminiscent of the covered wagon era. Bill rose and greeted his wife with a kiss.

"Stay away from me, I smell like one of Fat Jones' horses. I have been on the back lot all day doing a stage coach hold up scene."

"Never mind, you always smell like pear blossoms to me. Come

< 91 >

here." Bill put his arms around his wife's shoulders, and kissed her again gently. "What's for dinner?"

"It's sort of a surprise for Monday. I sent Mrs. Shafer to pick up a fresh rabbit from the Pepper Tree Rabbittree. I will start preparing it as soon as I get out of this outfit. Where's Billy?"

"He's in the backyard playing with Gypsy. Mrs. Shafer went home."

Feeney, swirling the ice in his high ball, followed his wife to their bedroom.

"Kathryn, I wrote a front page story for tomorrow's paper. They put the crank case oil kid on this story. Mary Olive Julian filed a divorce suit asking custody of Lois and Frances. She claims Julian is worth two million dollars and she is seeking temporary alimony of five thousand dollars a month."

"I know all about it Bill. She told me about her plans weeks ago. I kept quiet You told me never to pump information from her so I never shared it with you."

"That's what we agreed on, I don't fault you for that. The complaint is a mile long. The reporters took notes for hours. I sped up my research. I stopped by her attorney Leonard Wilson's office and obtained a copy of the complaint from him. She goes into a lot of detail about his love affairs."

"Bill, I told Mary Olive years ago that if you chased women like Julian, I would take our child, this home, and throw you out on the Cahuenga Pass for the red car to scoop up in its cow catcher."

"That's not a very nice thing to say. Why not just shoot me?" He laughed.

"Bill, it's not a matter to joke about. He treated her terribly. My friendship with her sustained her during some very trying times. That time he tried to have her declared insane to get control of her father's estate was just one of his hateful spiteful acts. Where is Julian anyway?"

"He's staying at the Hollywood Roosevelt Hotel with his body-guard, Tex Boren. The court ordered Julian to appear at Wilson's office at the Oviatt Building Friday at 2:00 p.m. for a deposition. Thus far, the process server can't find him. Is Mary Olive going to stay in Hollywood?"

"No, she hates the house he built. She's going to auction off the home and all the things he brought back from his trip to Europe, including the rare books. Also, he installed a gold bathtub which should bring something."

"Do you want to go to the auction, Kathryn?"

"No. Go if you wish. If I didn't know who owned the furnishings, I might bid, but I don't feel I should cash in on a friend's misery. Tell Billy to come in, it's getting dark. The wind is coming up. I can hear it off the top of Cahuenga Peak. Oh, how could I forget? Just as we suspected, Dr. Symons called the studio. He thinks nearly three months. I told you I was feeling funny. No periods. I love you, Bill and I love having our second child. I am happy for both of us and for Billy."

Bill Feeney, accompanied by Raul Dominguez joined hundreds of

< 92 >

persons who tramped through the nineteen rooms of the C. C. Julian mansion. The two men had signed their names and addresses on the roster of bidders and were assigned numbered paddles.

They entered the large living room paneled in beautiful softwoods. The auctioneer, George Fisher, assisted by A. N. Abell, coatless, shouted vigorously to be heard.

"Ladies and gentlemen, the contents of the large library which contains many rare volumes will be sold Friday at 2:00 p.m. You may view the volumes one hour before auction. Today, we will begin with fine tapestries, mirrors, tables, rugs, and fine pieces of furniture."

"Raul, let's get out of here. It's too hot. I'm only interested in some of the rare books. Let's turn in our paddles until Friday."

"O.K. by me, Bill. I don't have the bucks for this kind of stuff."

A continual stream of people passed through the upstairs bedrooms. They peeked out through the windows overlooking the city. In the rear portion of the property, children sat on the edge of the heart-shaped swimming pool, shoes and socks at their side, cooling their feet and splashing water on one another.

Julian's Rolls Royce, also for sale at auction, stood in the back driveway. At it's side, a soda pop vendor sold 5-cent pop for a dime to the thirsty crowd.

"Bill, there is your buddy Postal Inspector Madeira. He's coming our way."

"Are you covering this for the Citizen News, Feeney?"

"No. I don't cover celebrity auctions."

"I thought you might be covering the auction of a crook's ill-gotten goods, Feeney?"

"If they assigned me to such an auction, I would undoubtedly at this time be at the open market on North Western. They are auctioning the personal effects and belongings of Mrs. Guy Bates Post. She shot and killed her friend, Mrs. Doris Murray Palmer and then killed herself. I understand C. C. outsmarted you during your tax audit by quit-claiming this mansion to his wife and signing over title of the Rolls to Mary Olive. No liens, no beans."

"Yeah, Feeney, C. C. is real smart. This home cost him $250,000. Conservatively, the furnishings cost $100,000. It's just a matter of time, Feeney. Julian will slip up and I'll be there. See you around town, Feeney."

Chauncey Julian paced back and forth in his office at the Cotton Exchange Building in Oklahoma City. Squashing a cigarette in an ashtray littered with butts, Julian turned to Tex Boren.

"Call Riordon, tell him to get the plane ready. We're going to fly to Laredo, Texas. I need some cash and I think I know where I can get it. Also get me a snub nose .38. I may have to do some persuading. Have one of the salesmen tag along in case we need him."

The flight from Oklahoma City to Laredo, took fuel stops at Fort Worth and San Antonio. The Orion arrived in Laredo at sunset.

"Senor Julian!"

A small man with a dark mustache approached Julian.

< 93 >

"I'm Joe Flores. I've located your man. You'll find him at the Robert E. Lee Hotel. He's there every morning. Leaves about noon. Senor Julian, I made reservations for you and Tex across the border at Nuevo Laredo. Lots of night life." Flores approached Julian and whispered in his ear. "I got you the American passport—name is T. R. King. It's as perfect as the real thing."

"Thanks, Joe. Will a thousand dollars cover it?"

"Yes, Senor Julian. Will you need me tomorro?"

"No, Joe, we'll handle this ourselves. Tex, tell Clay and the pilot to take rooms at the Robert E. Lee Hotel. Tell Clay I want him to sit in the lobby at 9:00 a.m. tomorrow."

At 9:00 a.m., Julian and Tex Boren approached the registration clerk at the Robert E. Lee Hotel.

"Sir, I am C.C. Julian, President of C.C. Julian Oil & Royalties Company of Oklahoma City. My former accountant, Mr. Bolling, is staying with you here on business. May I have his room number?"

"Well certainly, Mr. Julian. He is in room 207. I'll ring him for you and tell him you are in the lobby."

"No, that won't be necessary." Julian placed a hundred dollar bill near the clerk's fountain pen holder. Julian called to his salesman.

"Clay, wait for us in the lobby."

Julian and Tex Boren climbed the stairs to the second floor. In a darkened hallway at room 207, Julian knocked lightly. The voice from within called, "Who is it?"

"Lamar, it's C.C. Julian. I need to have a word with you."

Julian motioned Tex Boren to step out of sight. Lamar Bolling opened the door. He was in shirt sleeves. Oil company promotional materials were spread out on his bed. Julian and Tex Boren entered the hotel room. Boren closed the door.

"Lamar, you made a lot of money with Julian Oil & Royalties the first three months of this year. Fact is, you got more than your share for preparing the promotional material. I was overly generous. Now, I need fifty thousand dollars, and I need it today."

"Are you crazy, C.C., I earned that money and I'm keeping it. How did you know I was staying here?"

"I have a lot of friends in Texas. Tex, pack his clothes, he's coming along with us. I'm going to get that money one way or another. Put you coat on, Lamar."

Julian withdrew the snub nose .38 from his right hand coat pocket motioning Bolling to the door. Tex Boren, after stuffing Bolling's belongings into his grip, followed Julian and Bolling down the hall to the staircase which led to the hotel lobby. Julian locked his left arm with the right arm of Bolling. With the .38 held under his coat he pressed the muzzle tightly into Bolling's rib cage. The three men formed a group at the cashier's desk.

"Mr. Bolling will be checking out."

"Of course, Mr. Julian, that will be thirty dollars."

As Lamar Bolling reached for his billfold in the upper left coat pocket, he caught a glimpse of a state highway patrolman standing behind

< 94 >

the group at the cashier's desk. Suddenly Bolling kicked backwards with his right foot striking the patrolman's left brown leather puttee.

"Officer, these men are kidnapping me. They have guns. This man has a gun stuck in my ribs."

Quickly the highway patrolman stepped back and ripped out his service revolver.

"All of you raise your hands slowly. You two, drop the guns on the desk. Now, all of you place your hands on your heads and keep looking forward. Gus, call Sheriff Condren. Get some deputies over here."

In a matter of minutes, sheriff's deputies arrived. They separated the three men, frisking all of them.

"Homer, what's going on here?"

"The fellow over there claims these two were kidnapping him. The tall fellow had this .38 in his right hand. Fatso had this .32 in his pocket. One of the deputies turned to Bolling.

"What's your name?"

"L.L. Bolling. I'm from San Antonio."

"Are you willing to file a complaint against these men?"

"I certainly am. The fat one is the bodyguard of this fellow C.C. Julian. He wanted me to give him fifty thousand dollars and when I refused, he threatened me and then they forced me out of my room at gun point.

Officer, that fellow sitting over there, he works for Julian. I think he's part of their mob."

"Mob, we're no mob. I am Chauncey C. Julian and these men are my employees. We came here to collect a debt from that welsher."

"All right, Julian, talk to the judge. You're under arrest. Let's go." The deputy handcuffed Julian and Boren.

"Homer, take that cream puff over there to the station in your patrol car. What's your name, boy?"

"Clay Mann, sir, I'm a salesman for..."

"Shut up. Cuff him, too, Homer."

"Hear ye, hear ye, hear ye. All rise. This justice court is now in session, Honorable Leopoldo Villegas, Justice of the Peace, presiding. Please be seated."

"Good morning, gentlemen."

"Good morning, your honor. John Valls for the people of the state of Texas."

"Good morning, your honor. M. J. Raymond representing the defendant C.C. Julian and C.C. Boren, both defendants are present."

"Mr. clerk read the charges."

"Yes, your honor. Count one charges that on or about April 2nd, 1931, in Webb County Texas, C.C. Julian (alias Herbert Murphy) and C.C. Boren (alias Tex Boren) did unlawfully detain L. S. Bolling against his consent by detaining the said L. S. Bolling with a firearm and threatening his life with intent to hold said L. S. Bolling for ransom and to extort money from him.

< 95 >

Your Honor, this charge was sworn to by Homer T. Sealey of Laredo Texas."

"Gentlemen, this matter will be referred to the Webb County Grand Jury. Bail is set at fifteen thousand dollars for each defendant."

"All rise, this honorable court is now in recess."

"Sheriff Condren, Julian and Boren have returned from court. Bail was set at fifteen thousand dollars."

"Art, lock up Boren and bring Julian to my office."

"C.C., get on in here. Shut that door. C.C., I told you yesterday you were in a peck of trouble. Webb County ain't no Hollywood. These are real folks here. You can't pull off this kind of stuff here. What got in your head?"

Chauncey Julian lit a cigarette. He paced back and forth in the Sheriff's private office.

"Look Joe, you've known me since the old days on the rigs. I've got a hundred dollar punch in my right hand, but I don't shoot people. I was just trying to frighten Bolling. He owes me some money. Things aren't going so well in Oklahoma. Folks up there don't appreciate me. The Indians got so much money, they don't need no investments and my mail campaign to California has fizzled. I'll work things out though, Joe, as soon as I make bail and thanks for your remembering me, old buddy."

"Stay in here C.C. until the bondsman comes through. I have to get over to the courthouse. And please C.C., don't come back down here to Laredo for a long, long time. That is, if you get yourself out of this mess."

"I'll say one thing Joe, you folks down here are honest men. Look at this front page. 'Al (scar face) Capone, hoodlum leader and beer king of Chicago smilingly walked from the court building here today surrounded by an escort of detectives—a free man. He was cleared of vagrancy by prosecutors who said they could find no policeman intimate enough with Capone's activities to testify against him. The action today freed Capone of all state charges against him and leaves him free to pursue any course he may choose. It is rumored that Capone hopes to break all connections with Chicago's rackets on a secluded ranch.' Joe, are you going to let him come down here to Webb County?"

"C.C., this isn't Chicago. Don't be silly. You're facing serious charges here. If that Grand Jury brings in a true bill on a kidnapping count, you could be facing the death penalty under a new Texas law. Matter of fact, if they do you will be the first person indicted under the new law as far as I know."

"The death penalty, come on Joe, I was just trying to frighten the guy. He owes me money."

"Look C.C., I don't make the laws, I enforce them. You can stay in my office until you make bail. See you later."

Chauncey Julian lit another cigarette and placed a call to his hotel.

"This is C.C. Julian. I'm registered under the name Herbert Murphy. Send my luggage over to the Hotel Condren. You never heard of the Hotel Condren? Its over near the courthouse. It has a star painted on the door. Think about it."

< 96 >

Julian hung up the phone and placed a second call.

"Operator, please get me the Robert E. Lee Hotel. May I speak to Lamar Bolling please?"

"Lamar, this is C.C., I hope you're feeling better. Lamar, how about going on the bond for me?"

"Going on the bond for you! After what you did to me! Certainly not!"

"Very well, Lamar, tomorrow I'm going to file suit against you in Oklahoma City for embezzling one hundred and twenty five thousand dollars from the C.C. Julian Oil & Royalties Company and you know who my witness will be. That fine young man who opened up the dummy account. Sleep well, Lamar."

After Julian posted bond, he met in his hotel room with his attorney.

"C.C., the Grand Jury named you in four specific indictments and a blanket indictment. Boren was named in three counts. Clay Mann was named in four complaints. The District Attorney would not stipulate as to what penalties he would be seeking. Let's get over to the County Jail and arrange for another bond."

"Wait a minute, Mike, before we go over there tell the reporters I want to talk to them in the lobby."

"That should be easy, there are several of them down there already."

Chauncey Julian entered the hotel lobby.

"Howdy, boys. Before I return to the Sheriff's office to arrange for another bond, I want to make a statement. I have contacted my attorney, O.A. Cargill in Oklahoma City. We are demanding that the County Attorney file a criminal complaint against Lamar S. Bolling, charging him with embezzling one hundred and twenty five thousand dollars which I gave Bolling last year to hold in trust for me. Bolling was my former publicity director for the C.C. Julian Oil & Royalties Company. He did a poor job and I discharged him, as is well known to my staff."

The following morning, Chauncey Julian received a phone call from the clerk of the court indicating that there would be a special hearing at 9:00 a.m.

"All rise, this court is now in session. District Judge J. J. Mullally presiding. Please come to order and be seated."

"Good morning, gentlemen.

"Good morning, your honor. John Valls representing the people of Texas."

"Good morning, your honor. Michael Raymond for the defendant Chauncey Julian who is present."

"Mr. Valls."

"Your Honor, I have requested that this matter be placed on calendar because of statements allegedly made by the defendant which reflect on the integrity of the court officials. Press reports all over the country state that Mr. Julian referred to the fifty thousand dollars he is alleged to have attempted to extort from the victim, L.S. Bolling, as 'chicken feed'."

< 97 >

"Mr. Julian, how do you respond to that charge sir?"

"Your Honor, I also read that statement and I assure you your Honor that it is false and without authority."

"Your Honor, if I may, due to the complexity of the charges against my client, I am requesting that these matters be continued. I've been busy with other cases as your Honor well knows."

"Your Honor."

"Yes, Mr. Valls."

"Your Honor, may we put this over for an hour so that counsel may confer. I assure you, your Honor, that Mr. Julian will be shown no favoritism because of the alleged statements while released on bond."

"Very well. We will be in recess for one hour."

"All rise, this court will be in recess for one hour."

"Court is again in session, please remain seated."

"Your Honor." "Yes, Mr. Valls."

"Your Honor, counsel had a conference. Present at the conference were myself, Robert Lee Babbitt, former Attorney General of Texas, Ed Mullally, Assistant District Attorney, and counsel for Mr. Julian, Mr. M. J. Ramond. Your Honor, I have agreed to a proposition that the defendant plead guilty to the charge of carrying an unlawful weapon and pay a fine of five hundred dollars."

"Very well, Mr. Julian please step forward. How do you plead?"

"I plead guilty."

"A plea of guilty is entered into the record. The defendant is to pay a fine of five hundred dollars."

"Your Honor, defense counsel and I will continue discussions concerning the disposition of the other counts."

"Very well, this matter will be continued to May 31st at 9:00 a.m. Any objections?"

"No, your Honor."

"No, your Honor."

"All rise, this honorable court is in recess until 11:00 am."

Chauncey Julian drained the last few drops from his coffee mug.

"Come on Tex, lets get over to the courthouse. Art, take a cab over to the airport. Wait for us there. Make sure the plane is refueled. Also order a picnic lunch for the flight back to Oklahoma City. Charlotte dear, here are a few sheckles. Do some shopping. Be back at the hotel by noon time. I want to get out of this burgh."

Chauncey Julian wearing a dark pin stripe suit, the same suit he wore at his first arraignment, sat inside the court rail with Tex Boren. Clay Mann, the third defendant entered the half filled court room along with the star state witness Lamar S. Bolling. The two men took seats at opposite corners at the rear of the courtroom. Judge J.J. Mullally ascended the steps to the bench and rapped the courtroom to order before the clerk could recite his usual call to order.

"Good morning, gentlemen."

< 98 >

"Good morning, your Honor. John Valls representing the people of Texas."

"Good morning, your Honor. Edward J. Mullally, also representing the people of Texas."

"Good morning, your Honor. M.J. Raymond representing the defendant Chauncey Julian who is present."

"Good morning, your Honor. Lewis J. Wardlow representing defendant Boren who is present."

"Good morning, your Honor. C.B. Neal representing defendant Mann who is present."

"Where is defendant Mann?"

"Seated near the rear of the courtroom your Honor. Mr. Mann will you please stand up?"

"Very well, you may proceed."

"Good morning, your Honor. Robert Lee Babbitt, Special Attorney representing the victim in this matter, Mr. Bolling, who is also present. Mr. Bolling will you please stand up."

"Are all parties ready to proceed?"

"Your Honor, may I be heard?"

"Mr. Valls."

"Your Honor, may we put this matter over to the second call? Your Honor we intend to prosecute the kidnapping charge under a new Texas law making the offense punishable by death. The minimum penalty is five years imprisonment. Your Honor, I've obtained a photostatic copy of the new law from Austin. It was signed by Governor Ron Sterling March second. I have made an extra copy for your Honor. May I approach the bench?"

"Certainly."

"Very well, we will call this matter at 11:00 a.m."

"All please rise, this honorable court will be in recess until 11:00 am."

District Attorney Valls led the attorneys into his office. The defendants, witnesses, and veniremen remained in the courtroom. At 10:00 a.m. District Attorney Valls entered the courtroom and announced, "All veniremen are excused until 11:00 a.m. All witnesses are to meet with my staff in the Grand Judge room on the third floor."

After the courtroom cleared, defense counsel met with the three defendants.

Judge Mullally reconvened the court at 11:00 a.m. District Attorney Valls rose, "May I be heard?"

"Mr. Valls."

"Your Honor, I am authorized to announce to the court that all differences that might have existed between Mr. Lamar S. Bolling and Mr. C.C. Julian have been satisfactorily adjusted. Mr. Bolling owes no money to Mr. Julian or to the Julian Oil & Royalties Company. Statements carried in the press and attributed to Mr. Julian which are derogatory to the affect that Mr. Bolling had embezzled one hundred thousand dollars or any other sum of money were not authorized by Mr. Julian and were not correct. Mr. Bolling, is what I've presented to the court accurate?"

< 99 >

"Yes."

"Mr. Boren?"

"Yes."

"Mr. Julian?"

Chauncey Julian slowly rose, stared for a moment at Attorney Valls, then responded "Yes sir."

"Mr. Julian and you Mr. Boren, is it your intention to plead guilty to assault with a prohibited weapon?"

Julian responded, "Right"; Boren responded, "Yes sir."

"Also, gentlemen, is it your intention to plead guilty to simple assault on a person of Mr. Junker, a witness?"

"Right," responded Julian; "Yes sir," responded Boren.

"And is it your intention Mr. Julian and that of Mr. Boren also to plead guilty to three other assault charges?"

"Right", responded Julian, "Yes sir", responded Boren.

"Your Honor, we ask your Honor to impose the maximum fines as to all counts."

"Very well, Mr. Julian and Mr. Boren, you will pay fines to the clerk in the sum of one thousand dollars as to the charge of aggravated assault and in the sum of five hundred dollars as to the other four assault charges. Anything further Mr. Valls?"

"Yes, your Honor, we move to dismiss all charges as to defendant Mann due to insufficient evidence. We also move to dismiss the charge of conspiracy and kidnapping as to defendants Julian and Boren."

"Very well, so ordered. The court will now be in recess."

"All please rise, this honorable court is now in recess."

"Tex, go over to the clerk and pay the fines."

Lamar Bolling quietly slipped out of the courtroom. The prosecutor met with reporters outside the courtroom.

"Well, gentlemen, I have no comment other than to state that within two or three days, I will ask Judge Mullally to reconvene the Grand Jury of the Forty Ninth District Court."

"Does it have to do with Julian Mr. Valls?" asked one of the reporters.

"No comment."

During the commotion, Chauncey Julian and Tex Boren slipped out of the courthouse taking a taxi cab back to the hotel.

"What's next, boss?"

"I wish I knew for sure, Tex. I've run out of gas in California and now Texas. If we can't make things go in Oklahoma, I may try China."

CHAPTER 15

United States Postal Director Madeira slouched in his swivel chair in his basement office at the Burbank, California Post Office. He was logging names, addresses, and dollar figures for losses as claimed in letters received from around the country, from various complaining postal patrons who

< 100 >

responded by mail to an advertisement placed in a popular pulp magazine claiming a cure for putrid stools. For the dollar they mailed to the post office box in Burbank, the postal patrons received a twenty-five cent bottle of castor oil. The renter of the post office box had not renewed the rental agreement since the last month's edition of the publication sold out at newsstands. The address he placed on the application for the box proved to be an unoccupied office building.

A bulletin board in Madeira's office, along with wanted posters, displayed a photograph of Chauncey Julian clipped from the Hollywood Citizen News front page story which covered the sensational details of the allegations of Mary Olive Julian as to her husband's habit of dallying.

The receptionist, whose cheerfulness had the effect of disrupting bureaucratic ennui, the result of routine law enforcement work, sang in to the loudspeaker "Inspector Madeira, a Mr. Penn to see you."

Inspector Madeira sat upright, buttoned his collar, adjusted his necktie, and combed his hair.

He then walked to the reception area where he was greeted by a nervous, rather tall man dressed like a store walker. He held a felt hat with both hands.

"Inspector Madeira, do you remember me? My name is H.A. Penn. I formally sold stock for C.C. Julian. You spoke to me several years ago when you were investigating complaints from investors."

"Certainly Mr. Penn, come into my office. Would you like some coffee?"

"No, thanks Inspector, it makes my ears ring."

"Well we don't want that now do we. Sit down please Mr. Penn. What's on your mind?"

"Inspector, I received in the mail a long letter and a financial statement from C.C. Julian. It was mailed from Oklahoma City, Oklahoma. He wants me to sell securities for him here in California. I remembered you asked me to cooperate with you in your investigation of Julian. I thought you might like to look at these documents. I have no intention of working for him. No one in Southern California would buy another Julian promotion and I am certainly not going to move to Oklahoma."

"Thanks, Mr. Penn, may I keep these? Where is the envelope these came in?"

"Here, Inspector, I left it in my pocket."

"Is this still your correct address?"

"Yes, Inspector."

"Do you have a phone number where you can be reached during daytime hours?"

"Yes, Inspector, here is my business card. I'm a floor walker at the Broadway Department Store in Hollywood."

"Thanks, Mr. Penn, I'll be in touch with you soon. Good day to you."

Madeira, now revitalized, carefully read the letter from Chauncey Julian to his former salesman.

"Listen to this, Charlie," Madeira called out to a fellow postal

< 101 >

inspector seated at a desk across the hallway, "This is from Chauncey Julian to a former employee."

"If you could be here and see with your own eyes the almost unbelievable facts—I question not for one moment but that you would reach out avidly, that's in Caps AVIDLY—feverishly to grasp to the fullest of your ability this fleeting fortune chance—for fortune chance it is—nothing less than a downright plunge for potential millions!"

"Now what do you think of that?"

"Horse hockey."

Madeira carefully examined line by line the financial statement of C.C. Julian Oil & Royalties Company and then placed a phone call to the Burbank Airport.

"When is your next contract air mail flight to Denver?"

"Sir, at 2:00 p.m."

"This is Postal Inspector Madeira. I'll be at the airport at 1:30. Reserve a seat for me. I have official business in Oklahoma City. I'll take an air mail carrier to Denver and transfer to the Oklahoma City Airline."

The Oklahoma City Airline pilot lowered altitude as the plane crossed over the North Canadian River. Winds buffeted the twin engine aircraft. It landed with dust puffing outward from the plane's wheels like a Model T on an Oklahoma highway. The plane taxied to a hanger where United States mail trucks were parked in a row. Inspector Madeira had wired ahead for a government automobile. It was parked along side the mail trucks. After displaying his government driver's license and identification papers, Madeira was given the car keys. He then drove south on North 39th Street toward downtown Oklahoma City. The State Capitol loomed in the distance. He turned left on Walker Avenue. His hotel was at the corner of Grand Avenue and Harvey Street. It was nearly midnight, too late to contact the U.S. Attorney's Office or the Postal Inspector's Office.

Madeira paid for his room, showered, changed clothes and left the hotel. He ordered dinner at an all night cafe near the Calcord Building. After eating, Madeira lit a cigar and walked toward the river down Walker Avenue. A babble from voices of a crowd in the middle of Walker Avenue grew louder as a group of men, nearly a hundred in number, shabbily dressed, some shoeless, were herded by nightstick wielding policemen toward the downtown area. Madeira stood on the sidewalk as the assemblage paraded by, a convolution of downcast men, goaded by angry looking policemen. Following the pitiful parade was a tall man in a dark business suit with a police badge pinned on his left lapel. He held a megaphone.

"What's this all about?," Inspector Madeira asked a passerby.

"Them's river bottom folks. Floaters. Used to be sharecroppers. Come the drought of 1930, they set up shacks down by the river. No work for them. Police Chief Watt says he's gonna get rid of them. Arrest them for vagrancy. There's a big fight going on. Watt locks them up and Old Alfalfa Bill Murrey, the Governor frees them with what he calls 'Cease and Desist' orders. Their kids get cared for through a soup kitchen run by the Veterans of Foreign Wars. They gives them stale bread and old canned fruit. Could be worse you know. Where are you from, sir?"

< 102 >

"Further west. I'm from a place where some folks as desperate as your river folks jump off a bridge."

The following morning Inspector Madeira drove to the United States Attorney's Office at the United States Courthouse.

"Is Roy St. Lewis in? Tell him, Postal Inspector Madeira from Los Angeles is here to see him."

"Just one minute, sir."

The receptionist in the glass enclosure placed a call to the U.S. attorney. She motioned to Inspector Madeira to enter a door, and activated an electronic device which unlocked the door. Inspector Madeira entered the hallway which lead to a door upon which a circular symbol of the Department of Justice was placed along with the title: United States Attorney. Madeira entered the office and was greeted with a handshake by U.S. Attorney Roy St. Lewis.

"Sit down, Inspector Madeira, what can I do for you?"

"I flew in from Los Angeles last night. I'm working on a possible postal fraud case involving the C.C. Julian Oil & Royalties Company here in Oklahoma City. I will be in town for a spell. I need to do more investigating. I know I have Julian on one count of mail fraud and possibly two. I have evidence which I'll present to you later after I complete my investigation."

"Yes, Inspector, I've been reading about C.C. Julian. There has been some civil litigation here in Oklahoma City both in the federal and state courts. I understand the company is in receivership. A local state court judge issued a restraining order forbidding Julian from disposing of any of the company's property until a final hearing is held. Last week the sister of a movie star, Enid Bennett, apparently a big investor, appeared in court on behalf of her brother and told the judge that Julian was unable to manage the affairs of the company because of his riotous living habits. As I understand it, Julian resigned from the Trusteeship of the company with the result that the receivership action was dropped. The newspapers listed a Charles W. Mason, supposedly a former Chief Justice of Oklahoma as the new trustee. The newspapers said that Mason appointed Julian to represent the company in Texas. Julian was quoted as saying something like 'Any of my enemies who think I am through are going to be fooled.' Inspector Madeira, are you one of Julian's enemies?"

"Mr. St. Lewis, I'm not his enemy. Let's put it this way, I've taken a personal interest in him. In Los Angeles he indirectly nearly destroyed the reputation of many of the leading bankers and business people. He allowed the now defunct Julian Petroleum Corporation to get into the hands of a couple of bunko artists. They enriched themselves through an overissue of stock and playing games with crooked local politicians. A former banker was killed in a Los Angeles courtroom by a half crazed Julian stock investor. The District Attorney out in Los Angeles went to prison for taking bribes. C.C. Julian, was not convicted of any crimes in California. However, he has no conscience when it comes to other people's money."

"Well, I have received no complaints here in Oklahoma City

< 103 >

concerning any federal law violations."

"That's why I'm here. To pursue the investigation of what I believe to be fraudulent financial statements which were mailed to California postal patrons from Oklahoma. I'll keep in touch with you. Where is Julian's office?"

"It's over in the Cotton Exchange Building. The newspaper mentioned that he lives on Hudson near Fifteenth in the fashionable home of a female friend."

"Is Tex Boren still with Julian?"

"I'm not sure. The papers mentioned Boren's name when Julian got into some kind of scrape in Laredo, Texas. By the way, Inspector, there's a fellow in town named Clay Mann. He was a salesman at the Julian Oil & Royalties last April. Julian flew him down to Laredo in his airplane to try to shake down a fellow named Lamar Bolling for fifty thousand dollars. All I know about the affair was what I read in the newspapers. Clay Mann almost went to prison, along with Julian and Boren. Apparently the District Attorney in Laredo gave Mann a walk and recommended a fine for Julian and Boren. Originally, there was a kidnapping count against Julian and Boren which was dismissed for some mysterious reason. The man may cooperate with you. I believe he's still in town."

"Thanks, Mr. St. Lewis."

"Call me, Roy. Stop by the office for a cup of coffee anytime."

"I certainly will, Roy. Call me Manny."

Clay Mann was not difficult to locate. He was listed in the phone book. At 7:00 a.m. the next morning, Madeira arrived at Clay Mann's residence, an unpretentious, yet not inexpensive, home near Central High School at the corner of Robinson and Seventh.

"Good morning, sir, are you Clay Mann?"

"Yes, sir."

Madeira withdrew a leather case from his coat pocket, unfolding it revealing a United States Postal Inspector identification card on the inside and on the outside a Postal Inspector's silver badge.

"I am Inspector Madeira, United States Postal Inspection Service. May I have a few words with you? I am proceeding with an investigation of Mr. C.C. Julian. I believe he was your former or is he still your employer?"

"Former, I'm selling cars now. I stopped trying to sell oil stock when the production of oil in the East Texas fields drove the price of a barrel down to ten cents. Come on in, Inspector, the wife and children are still asleep. Would you like some coffee?"

"No, thanks, I had breakfast at the hotel. Clay, may I call you Clay?"

"Sure.

"Clay, I want you to understand that you are not under investigation by the Postal Inspection Service. I understand you were arrested along with Julian in Laredo, Texas."

"Yes, the charges were dismissed. I didn't do anything except sit there in the lobby like a dummy."

< 104 >

"We're not interested in what went on in Texas. But we are concerned with the violation of federal law, mail fraud. I have here a copy of a financial statement which is supposed to represent the assets of C. C. Julian Oil & Royalties. Would you look at it?"

"Surely."

"Is this a financial statement which Julian mailed to his potential investors?"

"Yes, it is."

"Did you have anything to do with financial matters at Julian's company?"

"Yes, shortly after I went to work for him, in 1929, December as I recall, Julian asked me to handle his banking in my name at Fidelity National Bank. Each day I would deposit checks and some cash from the sale of stock. I would then obtain cashier checks made out to myself, fifty percent of the proceeds in one check and fifty percent in another. Then, I would endorse the checks, giving one to Mr. Julian and the other to Mr. Bolling. Mr. Bolling was in charge of public relations at that time."

"What did they do with the money?"

"I really don't know."

"Did you profit from the arrangement?"

"Well, Julian promised to pay me two and one half percent, but he never did. The only money I ever made was from commissions on the sale of stock."

"Look at this financial statement; it shows the C.C. Julian Oil & Royalties Company with $418,267.18 cash in the bank on January 31, 1931. Is that true?"

"As far as I know, Mr. Julian had no bank account for the company."

"Who prepared this financial statement?"

"As far as I know, Leo Young, the bookkeeper, and Mr. Julian."

"Is Leo Young around?"

"Yes, he lives down the street. His phone number is listed in the phone book."

"I'll give him a ring later. Clay, you've been very helpful. I would ask you not to discuss this interview with anyone. I don't see any need for you to retain an attorney as we have no intention of indicting you. Just stick around town. We will be asking you to testify at a bail hearing."

Later, after a brief interview with Leo Young who promised he would cooperate, Madeira phoned the United States Attorney's office, arranging for an appointment at 8:00 a.m. on Monday, July 13th.

"Good morning, Roy."

"Good morning, Manny, care for some coffee?"

"Yes, black. Thanks. Listen Roy, I'm ready to proceed with the Julian matter. I've got two good witnesses who will testify as to the false statements in the financial report which was mailed to former sales people in California. I can have my two witnesses who are on call flown here to Oklahoma City by mail planes, one from Burbank, and the other from Oakland. I can be ready by Wednesday morning, if you give me the go ahead."

< 105 >

"Do you have the material here with you that was sent out by mail?"

"Yes, I'll give it to you now."

Madeira reached into his briefcase and took out two identical letters, financial statements and two envelopes with stamps cancelled in Oklahoma City.

"These were mailed to postal patrons in California."

"All right, Manny, wait here and I'll have some complaints prepared charging mail fraud. Let me see, we will allege that more than two million dollars was obtained through the sale of trust certificates in whole or in part by the fraudulent use of mail. I'll have warrants prepared. Why don't you come back after lunch and pick up the warrants. I'll ask the United States Marshals to have some deputies available to assist you when you arrest Julian. Meet them in the parking lot at 4:00 p.m."

At 4:00 p.m., Inspector Madeira met two Deputy U.S. Marshals in the parking lot of the United States Courthouse. The three men entered a small bus used by the Marshals to transport prisoners and drove off to Julian's office at the Cotton Exchange Building. Madeira, without knocking, opened the door to Julian's office. He was followed by the two Marshals. The three federal agents had affixed their badges by hooking them to the exterior of their coat pockets.

"Chauncey Julian, you are under arrest for mail fraud. Tex, keep your hands on the table. You are also under arrest. Who are you?"

A third man seated at the opposite side of the room replied "I am H.D. Topp, Mr. Julian's accountant."

"You are also under arrest. Each of you place your hands behind your backs. Marshals, cuff them."

"The two Marshals placed handcuffs on the three men.

"Pat down old Tex there. See if he is packing."

"Nothing, Inspector."

Julian smiled. "Manny, what are you doing in Oklahoma? I haven't seen much of you since I sued your butt."

"I have never forgotten you, C.C.. Save your talk for your attorney. Let's go."

"What's the bail on your warrant, Manny? I assume you have one."

"Fifty thousand on your warrant C.C. and ten thousand on old Tex here and Mr. Topp."

"You'll never get away with this Manny, you're out of your element in Oklahoma."

"C.C., you're out of your element no matter where you go. C.C., you're the biggest four flusher in the USA as far as I'm concerned. Now, shut your mouth and let's go."

Julian, Boren, and Topp were transported to the United States Courthouse where they entered pleas of not guilty before a United States Magistrate . Failing to make bond, they were booked and lodged at the county jail.

Wednesday morning, 9:03 a.m., United States District Court, Oklahoma City.

< 106 >

"All please rise. Here before the flag of our country, this honorable court is now in session. The honorable George Eacock, United States Commissioner presiding. Please be seated."

The cryer then struck his gavel and all present were seated with the exception of the attorneys and over twelve hundred people who crowded in and around the small courtroom, filling the halls and drifting down to the front steps of the courthouse.

"Good morning, your Honor, Roy St. Lewis representing the government."

"Good morning, your Honor, J.B. Dudley for defendants Topp and Boren who are present."

"Good morning, your Honor, O.A. Cargill for defendant Julian who is present."

"Gentlemen, the purpose for this preliminary session is to determine whether there is sufficient evidence to bind these men over to the United States District Court on the charge of mail fraud. Mr. St. Lewis, you may proceed."

"Thank you, your Honor."

"The government calls Mr. Clay Mann."

"Mr. Clerk, please swear in the witness."

"Do you solemnly swear to tell the truth, the whole truth and nothing but the truth so help you God?"

"I do."

"Your name, sir, for the record."

"Clay Mann."

"You may proceed, Mr. St. Lewis."

"Thank you, your Honor. Mr. Mann, where is your residence?"

"Currently, Oklahoma City. I'm originally from San Antonio, Texas."

"Were you formally an employee at the C.C. Julian Oil & Royalties Company?"

"Yes, sir, I was employed by Mr. Julian in, I believe, December of 1929 as a salesman."

"Did you have any other duties other than sales of securities?"

"Yes, sir, each day I converted checks received from investors through my account at Fidelity National Bank into two cashier checks made out to myself. I would then give them, after I endorsed them, one to Mr. Julian and one to Mr. Bolling, the publicity man."

"Your Honor, as government's Exhibit One, I wish to enter into evidence this financial statement which was placed in a mail depository for delivery to a postal patron in Burbank, California."

"So ordered. Mr. Clerk, please mark the document as government Exhibit One."

"Now, Mr. Mann, this document states that in January 1931, C.C. Julian Oil & Royalties Company had cash on hand in the bank in the sum of $418,267.18. Is that a true statement?"

"I believe not. As far as I know Mr. Julian had no bank account either for himself or for the company."

< 107 >

"Also, Mr. Mann you will note that the financial statement indicates as an asset bonds valued at $261,396.29. Is that a correct statement?"

"Well, sir, I saw the bonds and they were issued for Mr. Julian's California Oil and Mining Companies."

"Now, Mr. Mann, did you ever attempt to borrow money from Mr. Julian?"

"Yes, sir. Last December we couldn't sell any royalties shares. The price of crude dropped due to the over-production of oil in Texas. I asked Mr. Julian for a loan of a thousand dollars. He told me he was broke. He then asked me to go to Laredo, Texas with him to borrow fifty thousand dollars from Mr. Bolling. Mr. Julian was arrested for attempting to kidnap Mr. Boiling and extort the fifty thousand dollars. However, those charges were dismissed."

"Now, Mr. Mann, apparently some dividends were paid to investors. How did that transpire?"

"Mr. Julian used the money he received from the sale of stock."

"Were there any other officers at the C.C. Oil & Royalties Company who were responsible for the preparation of this financial statement?"

"As far as I knew, there were no officers or directors of the company except Mr. Julian."

"Sir, were you to receive any remuneration for the handling of monies in the way you described?"

"Yes, sir, I was supposed to receive two and one half percent for handling the bank accounts, but I never received a dime except commissions from the sale of stock."

"No further questions."

"Your Honor."

"Mr. Cargill."

"Just one question, your Honor. Sir, is it not true that the bank account arrangement was the result of a law suit involving a former account at the First National Bank & Trust Company?"

"I am not aware of that, sir."

"Any further questions? You may step down. Next witness."

"Your Honor, the government calls Mr. Leo E. Young"

"Mr. Clerk swear in the witness."

"Mr. Young, what was your position at the C.C. Julian Oil & Royalties Company in January of this year?"

"I was Assistant Bookkeeper."

"Sir, I show you the government's Exhibit One. Do you recognize this document?"

"Yes, sir, I helped prepare it."

"Thank you, sir, no further questions."

"You may step down. Next witness."

"Your Honor, the government calls Mr. Penn."

"Mr. Clerk, swear in the witness."

"Mr. Penn, where is your residence?"

"Burbank, California."

"Are you a former employee of Mr. Julian?"

< 108 >

"Yes, sir, I worked as a salesman for him in California."

"I show you the government's Exhibit One. Have you seen this document before?"

"Yes, I received it in the mail along with a letter from Mr. C.C. Julian."

"Thank you, nothing further. Your Honor, the government has no further witnesses."

"This matter will be bound over to the District Court after the matter is presented to the Federal Grand Jury in September."

"Your Honor, may I be heard as to the reduction of bond?"

"Yes, Mr. Cargill."

"Your Honor, in the discussion we had before this hearing you indicated you would be willing to reduce the bond if Mr. St. Lewis would so recommend."

"Mr. St. Lewis?"

"Your Honor, the government opposes any reduction of the bond. Mr. Julian is a British subject. He was born in Canada and I believe he is a flight risk."

"Very well, since these fellows are aliens, there will be no reduction of bail. This court is in recess."

The cryer pounded his gavel, "All rise. This court is now in recess."

The United States Marshals approached Julian, Boren and Topp, placing handcuffs on each defendant's hands clasped in front rather than behind. A group of reporters moved near Julian.

"C.C., are you going to make bail?"

"I couldn't make the bail if it was cut to five thousand dollars. I haven't got a thing."

"What are you going to do C.C.?" a second reporter called out.

"I don't know, stay in jail until September I suppose."

As Julian and the two other prisoners were escorted from the courtroom by the Marshals, Julian called out to the prosecutor, "Hey Roy, come over to my cell at the county jail and play pinochle with me. This fellow Boren can't play anything."

CHAPTER 16

A heavy cold, wind driven rain beat a staccato on the large plate glass window of Mary Olive Julian's home, west of Assiniboine Forest. The house was a portion of Mary Olive's inheritance from her father's estate. It was the last week in April; Lois, a senior, and Frances, a junior, had returned to school after the Easter recess.

Winnipeg, the capitol of Manitoba was a welcome relief for Mary Olive after the turmoil created by her husband who artfully avoided process servers in California leaving no resolution as to her divorce action in the Los Angeles Superior Court. Her life was now more serene. The girl's grades had improved in the Canadian public schools after the transfer from the exclusive girl's school in Los Angeles. Mary Olive was completing a letter to her

< 109 >

solicitor, who had prepared a lawsuit for argument before a local magistrate, relating to a neighbor's tree which, during a wind storm fell into her yard partially demolishing her fence. The neighbor had refused to haul the tree away or pay for the repair of the fence. The doorbell rang. The housekeeper entered the living room.

"Mrs. Julian, your husband wishes to speak to you. He is in the parlor."

Mary Olive drew her hands to her lips, a small white lace kerchief covered their trembling. She then walked over to a small mirror on the wall and composed herself. Momentarily, she opened the French door leading to the parlor.

Chauncey Julian rose from his chair as his wife entered the room. He was attired in his usual dark pinstripe suit, vest, white shirt with french cuffs, and a silk tie. The Brooks Brothers symbol was embroidered in dark blue on the tie. He held a black fedora in one hand and gold knobbed cane in the other. A heavy navy blue overcoat was draped over the back of his chair. Mary Olive closed the French door.

"Good morning, Mary Olive."

"What do you want, Courtney?"

"Aren't you going to ask me to sit down, Mary Olive?"

"I wasn't, but you may if you wish."

Julian sat down placing his hat on an end table and, retaining the cane with both hands, slowly massaging the gold knob. Mary Olive walked to a window staring at the downed tree in her side yard.

"Why did you come here?"

"I've come upon hard times, Mary Olive."

"You mean the Hoover depression put an end to your schemes, or there are no fools left in the United States to pay for your excesses."

"Neither is true! You know the destruction of Julian Petroleum in Los Angeles was not my fault. When I entered the Oklahoma City gusher field the politicians tried to tie me down with their stupid unconstitutional proration orders and conservation laws. If I had not been forced to spend enormous sums of money fighting the politicians and the courts, I could have made it big in Oklahoma."

"Courtney, you always blame someone else for your failures. If it's not the Corporation Commissions, it's the Mayor or Governor, or the courts."

"Mary Olive, I don't like people telling me how to run my business affairs."

"Courtney, did you ever stop to think that the Corporation Commissions of California and Oklahoma may have had some concern for your investors?" Mary Olive seated herself at an oak desk and turned to her husband. "I read in the paper that your bail was reduced by the United States District Court Judge in Oklahoma City to twenty five thousand dollars and that it was posted with surety of a pledge of a woman's home. Are you now a gigolo as well as a philanderer?"

"Not at all, my attorney's wife, Mrs. O. A. Cargill, pledged the loan and signed the bond as did Frank Russell and John Peacock, oil men who

< 110 >

owed me money."

"Do the conditions of your bond in Oklahoma allow you to enter a foreign country?"

"No, but I have no intention of returning to Oklahoma or the United States."

"Don't tell me that you have come home to be a loving husband and caring father for if you dare to try that line on me, I will simply call the police and have you arrested as a fugitive."

"No, Mary Olive, that sort of life never fit in my plans. However, I am not finished. I'm going to the Orient."

"To the Orient. With whom, your attorney's wife?"

"No, Mary Olive, by myself. I'm going to Shanghai to mount a campaign to develop the petroleum industry in China. I've been doing considerable reading about the Orient."

"Courtney, you know nothing about the Orient. You could not become an expert on the Orient if you remained in jail where you belong and read books day and night. Really, Courtney, why are you here?"

"Well, to tell the truth Mary Olive, I'm broke. I need some money to get to the Orient and start Julian Petroleum one more time. I think the Orient is ready for the development of its own petroleum industry."

Mary Olive shook her head, rose from her chair, walked to the French Door, opened it and called out "Adrian, please bring us some tea!" She closed the door and stood facing Chauncey Julian. "Very well, Courtney, you have the illusion you are a businessman, I will offer you a deal. I sold your Rolls Royce for ten thousand dollars and the gold plated bathtub in the shape of a heart for five thousand dollars. That makes a total of fifteen thousand dollars. I will write a check and have my bank prepare that amount for you in American dollars at twelve noon today under one condition."

"And what is that Mary Olive?"

The housekeeper opened the French door, pushing in a tea dolly. Mary Olive poured tea.

"Cream and sugar, Courtney?"

"Yes, thank you. The condition, Mary Olive?"

"The condition is that you leave Winnipeg on the Canadian Pacific Railway today for Vancouver. You are to have no contact with my daughters."

"Well, Mary Olive, its been over a year since I last saw them in Hollywood. I believe we had lunch in the Roosevelt Hotel. May I see a photo?"

"There on the piano."

Chauncey Julian picked up a gold framed photo of his two teenage daughters and then replaced it gently on the scarf which covered the top of the grand piano.

"Also, I don't want you to start writing letters to them. Just keep your distance as you usually do. These girls have a life here before them. You ruined my life, I won't let you ruin theirs."

"Very well, Mary Olive, its a deal."

She opened her check book and wrote a check.

< 111 >

"Here Courtney, take this to my bank. The address is on the check. I will call them and give my approval to cash it, $15,000 American. Good bye, Courtney."

"Good bye, Mary Olive."

Mary Olive remained seated as Julian put on his overcoat. He opened the French door to the entryway. Mary Olive called out, "Adrian, will you show my husband to the door."

It had stopped raining. Julian entered the cab which remained in front of Mary Olive's home. He was driven first to the bank in the Exchange District, and then to the railway station of the Canadian Pacific off Main Street. He purchased a one way ticket to Vancouver, British Columbia, and arranged for his several pieces of luggage to be shipped to Vancouver. Julian then entered a small restaurant on York Avenue. He removed a writing pad from his briefcase and wrote the following letter:

Mr. Walter M. Harrison, Managing Editor:
Oklahoma City Times, Oklahoma,

January 31, 1932,
Vancouver, British Columbia.

Dear Sir,

During my efforts to develop the holdings of the C.C. Julian Oil & Royalties Company, the newspaper you control agreed with the Oklahoma State Corporation Commission which brought suit against me because I defied the unconstitutional Oklahoma proration and conservation laws which prevented me from sharing in the world's wonder field of fields at Oklahoma City. We had locations for many great wells to be drilled. Our number one well was being prepared to set the casing at five thousand six hundred feet. We were drilling our number two and had made intensive preparations to move on our number three and number four. I wanted to drill in those great wells one right after the other just as fast as all the power of men and steam could drill them. I was just about ready to make good with a vengeance—wells one right after the other just as fast as all the power of men and steam could drill them. I was just about ready to make good with a vengeance to vindicate my honor—pay back all that my investors lost and plenty more. The polititians applied to me your unconstitutional conservation   law; yet, the wells of their friends and backers were producing one hundred percent capacity with no attempt to control drilling.

Well, you won't see me again! I am jumping bail and for the present will remain in Canada. Check your law books, the extradition treaty between the United States and Canada does not cover mail fraud cases.

Earnestly yours, C.C. Julian.

< 112 >

Julian placed the letter in an envelope, sealed it affixing a United States stamp. He then remembered he was in Canada. He walked to the front of the train station and on Main Street noticed a motor vehicle with an Illinois license plate. He approached the driver.

"Hi partner, here is a buck. Mail this letter when you get home. I don't have any Canadian stamps."

"Sure, I'm driving straight through to Chicago. I'll mail it sometime tomorrow in Chicago."

"Thanks, partner."

Chauncey Julian returned to the railroad station awaiting the train for Vancouver. He again took out his writing pad from his briefcase and printed boldly, "What price Fugitive? The memoirs of C.C. Julian."

Chauncey Julian rose from his assigned seat in the dining roon of the President Jackson. He had boarded the ship in Yokohama bound for Shanghai, arrival date March 23, 1933.
Julian utilized his falsified United States passport with the name E.R. King which was prepared for him in Nuevo, Laredo, Mexico during his visit in 1931.

"Dr. Wang, it was a pleasure dining with you these several days."

"It was my pleasure, Mr. King. As to your future plans, the actual and potential mineral wealth of China, before the present world depression, is about 2.3% of that of the entire world, and 4.7% of the mineral wealth of United States. Mr. King, I suggest you interest yourself in our exportable surplus of minerals such as tungsten or tin rather than oil. Or you may wish to interest yourself in silver trading. Sir, you may contact me at the brokerage firm of Benjamin & Potts in Shanghai. Here is my business card. Again it was a pleasure to have dined with you, Mr. King."

"Indeed sir, the pleasure was all mine."

The two men bowed to one another and Julian walked out of the dining area onto the first class deck. As the ship proceeded up the brownish water of the Whang Poo River, Julian could see oil supply depots, with their silver tanks. the property of foreign firms. The ship then passed the Shanghai Power Company plant while numerous junks dodged the bow. In the distance through haze, the skyline of Shanghai with its eight and ten story buildings emerged. An attractive young woman with long raven hair gently locked arms with Julian snuggling up to him.

"Mr. King, remember you promised me I could work for you as your secretary here in Shanghai. You won't renege, will you?"

"Of course not, Leonora. You say that your brother works for a financial house?"

"Yes, Mr. King. I will introduce him to you."

As the President Jackson approached its berth, the couple stood arm in arm. Off the bow of the ship were moored huge wooden Ningpo Junks, their sterns decorated with colorful variations of the Phoenix, the emblem of immortality. Crews unloaded large wooden poles, apparently for building purposes. Secure at her mooring, the passengers of the Dollar Liner President Jackson disembarked.

< 113 >

A group of reporters and photographers singled out one passenger for photos and an interview. Chauncey Julian had met the man aboard ship, a Mr. Bruce Barton, author of "The Man Nobody Knows" and more recently the head of an advertising agency in New York who with his wife and daughter were on a trip around the world. Julian sought his advice concerning publication of his book without indicating the title, and had been provided with the name and address of a publishing house in New York. The author had not been unfriendly to Julian and unenthused about Julian's project.

On the dock Leonora Levy ran a few steps throwing her arms around a young man who kissed her on the cheek and hugged her.

"Mr. King, come meet my brother, Ralph. Ralph, this is Mr. T.R. King, an oil man from the United States. He has come to Shanghai to explore opportunities in the oil business."

"Happy to meet you Mr. King."

"Where will you be staying, Mr. King. I would be happy to drop you by your hotel. My car is parked over there," gesturing toward the roadway along the Embarcadero.

"That is very kind of you. I'll be staying at the Cathay Hotel on the Bund."

"Mr. King, you'll have to make arrangements at the Customs House on the Bund to have your luggage delivered at the hotel."

Ralph Levy led his sister who had been vacationing for a month in Japan and Julian to his automobile, a Ford sedan. With Ralph and Julian in the front and Leonora in the rear, Ralph drove from the port on the Whang Poo River to the Zia Zi Road making a half circle around the Chinese city. He then turned left at the direction of a tall Sikh traffic policeman and through a maze of electric tramcars, motorbuses, motorcars, wheelbarrows, and Coolies carrying loads of incredible weight, he drove past the race course, turning right on Nangking Road, he parked in front of the Cathay on the Bund.

"Thanks for the ride, Ralph. Leonora, call me the first of next week and we'll discuss your secretarial duties. Thanks again for your help, Ralph. Bye bye, my dear."

"I'll call you on the third, Mr. King."

"Thank you, my dear."

Julian entered the hotel, one of several in the international settlement of Shanghai.

"Do you have my reservation, sir, Mr. T.R. King."

"Yes sir," responded the tall elegantly dressed Englishman at the reservation counter. "We will arrange to have your luggage transferred from Customs if you wish, Mr. King."

"That will be fine."

"Here are your keys. Front. Show Mr. King to his room."

Julian had purchased a copy of REVIEW, an English language weekly in the lobby. He took off his coat and tie, sat down and began thumbing through the magazine, looking for possible leads for his venture. He found a page called "Men and Events" and began underlining possible

< 114 >

contacts. There was a knock at the door. Julian opened it and uniformed Chinese bell hops brought in his luggage and carefully unpacked his clothing, placing his garments in the closet and chests of drawers. Julian gave each a two dollar bill, American. After a bath and a shave, Julian dressed, and cane in hand, left the hotel crossing Nang King Road, dodging rickshaws and tram cars.

He walked past the Bank of Communication and the Central Bank of China. Next he passed the Hong Kong and Shanghai Banking Corporation, with two massive bronze lions on either side of its locked gate. Finally, Julian entered the Shanghai Club, the home of the world's longest bar. He walked along the crowded bar until he caught sight of a businessman, obviously an American. Julian elbowed in next to the man, ordering a Bourbon and soda.

"First night in Shanghai sir?" asked the businessman.

"Yes sir, T.R. King. I'm in oil investments. And you sir?"

"I'm J.F. Malone. I represent American interests purchasing silver. What do you intend to do in petroleum here, Mr. King?"

"I believe there are oil reserves here and I hope to develop them."

"King old boy, I wish you well, but this is a silver boom town, not an oil boom town. The people in the interior of China are too poor to have machines that burn petroleum. The oil they sell here is for the power plants and the military. The deals are all controlled by the Chinese government at Nangking. Silver is the only thing the Shanghai businessmen and bankers are interested in. There are nearly six hundred million Mex-dollars worth of silver in Shanghai right now. Did you see all those banks out on the Bund. The shroffs work night and day in their vaults counting Chinese dollars. The Sun Yat-Sen dollar, the Dragon dollar, the Yuan Shih-Kai dollar, and chinese eagle dollars. Look King, there are four hundred million human beings in China; there are four million human beings in Shanghai. China is poor and Shanghai is rich."

Malone took a sip from his drink.

"Is that so, Mr. Malone? Bartender, another drink. Well perhaps I'll get into stock speculation."

"That's a possibility, King. Listen, stop by and see me when you get situated. I'm staying at the Palace Hotel. I've got to run now. Nice talking to you, King."

"Yes, of course, Malone."

Julian sipped his drink and dejectedly placed a two dollar bill to cover his bar tab. As he rose to leave, a woman who quietly seated herself next to him during his conversation with Malone tapped Julian on the shoulder. Julian turned abruptly.

"Mr. King, I couldn't help but hear your conversation with the American. I am Mary Cantorovich, formally from Moscow. I would very much like to introduce you to some of my friends."

"Why certainly, my dear, right now?"

"Not right now, Mr. King, but let's get out of here, it's so noisy."

"Your place or mine?"

"Mine, Mr. King. Mr. King, you see that tall Chinese bartender.

< 115 >

There in front of the mirror. Go purchase a bottle of champagne. We will take it to my apartment."

Julian returned to the bar. "A bottle of champagne! Whatever you recommend."

"Certainly, Sir. That will be two hundred American dollars."

"Two hundred dollars, are you..." Julian turned to the woman. She was seated at a small round top table. "I'm being set up. You are nothing but a God-damned whore."

"Mr. King. I am a Russian Countess."

"You are a God-damned Royal frigging whore. I never pay women. Women come to me. If I like them, I reward them."

"Don't you like me?"

"Come to think of it, no! You are too fat. Go over and trick one of those bloated Chinese bankers. I'm getting out of here."

Dinner at six—the usual table—Chauncey Julian and Leonora Levy. The Park Hotel across Nang King Road from the race course. The menu in French. Julian translates for Leonora.

"Francoise, a bottle of Mums."

Suddenly a woman's voice almost shouting startled Julian and drew the attention of other diners.

"Chauncey Julian. Whatever are you doing in Shanghai?"

Winnie St. Cyr approached Julian's table. Julian chagrined, arose.

"Why Chauncey, is this your daughter?"

"Miss Levy, this is Miss St. Cyr."

Leonora Levy appeared dismayed.

"T.R., why does she call you, Chauncey?"

"It's a long story Leonora. We'll discuss it later."

"Chauncey, I hear that they've been looking all over Canada for you! That nasty old bail bondsman. He even came to my apartment in Hollywood. Remember the cowboy? Its too bad I'm leaving tomorrow Chauncey, I would just love to talk about old times. Remember the Rose Parade? My dear, give old Chauncey everything he needs and he will be good to you. Give him a hot toddy before he goes to sleep and he won't wake up until morning. It will be a lot easier on you. Lestyle, C'est L'homme. You remember French, don't you Chauncey. Bye bye."

"I thought you were, Mr. T.R. King."

"My dear, my real name is Chauncey Julian. Don't worry, I'll explain everything to you after dinner."

"Excuse me, Sir."

A young man seated nearby rose and approached Julian's table.

"My naine is John Cookson. I'm with the New York Times. You are the Chauncey Julian! Are you not?"

"Yes sir."

"May I join you?"

"Well, you already have. Sit down. This is my secretary, Leonora Levy."

< 116 >

"Pleased to meet you, Miss Levy. Could you bring me up to date Mr. Julian?"

Julian reviewed the history of Julian Pete in California and the C.C. Julian Oil & Royalties Company in Oklahoma. Leonora Levy listened awestruck.

"John, I have studied data carefully which has convinced me that there are significant oil resources here in China. If one may go on the law of averages, I surely am entitled to a better break in China than I received in the United States. I am not confronted here in China with the powerful influence of enemies such as those California people who created the over issue of stock in Julian Petroleum after my stepping down as president. That group of bankers and politicians, including the so-called receivers caused a stigma to be attached to name which wrecked my business deals in Oklahoma. It was the pressure of powerful hidden forces that compelled me to dispose of Julian Petroleum Company and then the Merger Mines."

"Mr. Julian, what precisely went wrong in Oklahoma leading to your indictment by the federal government for the use of mails to defraud?"

"Well John, in a nutshell, the state of Oklahoma shut down production and the federal government indicted me because I could not pay the dividends as promised. In China I hope for better things and a square deal."

"Thank you very much, Mr. Julian. I hope you enjoy your dinner. It was a pleasure to have met you, Miss Levy."

Leonora Levy placed her hand on Julian's.

"What should I call you?"

Julian appeared puzzled. "Not Chauncey—too many bad memories. Certainly, not Courtney. There is only one person who calls me Courtney. C.C. dear, call me C.C."

"C.C., I love you."

"Leonora, there may not be time left for love. Love is a long term deal. None of my deals have been long term."

"C.C., everything in China is long term. That's why I live here. In the United States it's two years for this, four years for that, so many years until you retire, twenty one years to vote, clock in, clock out. C.C., don't rush things. You're in China."

"Leonora, I've made a discovery here in Shanghai and it frightens me. All my life my successes have depended on communication with the people—the little guys—the common folks—the hoipolloi. I can't get through to them here. I can't communicate and what is more, the little guys here haven't got anything. Shanghai is a lot like Los Angeles. It's run by bankers who are milking China dry of its silver. The bankers don't care about developing the resources of China. They are greedy, just like the big shots in Los Angeles."

"Keep trying, C.C. I'll be at your side as long as you want me. I told you, I love you."

"I believe you, Leonora."

"I have an appointment Monday with two Directors of the Chinese Development Corporation Leonora. I'm going to try to interest them in retaining me to negotiate with American businessmen in stock trading. I met

< 117 >

a Major Edward Howard at the Shanghai Club. He represents Douglas Aircraft Company and its subsidiary Northrop Company. He just opened an office in the Dollar Building. He told me that his company has long been a contractor to the Chinese government. He sells the Chinese light bombers and observation aircraft. He claims Northrop Company manufactures the most advanced allmetal aircraft. I asked him if he would be on the lookout for a position for a man with my background. He laughed at me. He said he has a framed Julian Pete stock certificate back in the States. Then he bought me drink."

Later, Julian and Leonora were strolling on the Bund. At the foot of the statue of Sir Robert Hart, a news vendor was hawking the new edition of the English language REVIEW. Julian took some change from his pocket, buying a copy. The headline on the front page with the date line December 23, 1933 was "Japanese Invade Chahar—Marshall Chang Hsueh-Liang Returns to China." Julian glanced at the list of articles in the issue.

"Leonora, look at this."

"Insull must leave Greece, but Shanghai protects Julian."

Julian flipped pages until he found an article on page 138:

"Samuel Insull, Chicago Promoter, sold the American people some $3,000,000,000 worth of securities in public utility enterprises, much of which turned out to be worthless. In order to escape prosecution he fled to Greece where the United States has no extradition treaty. The American Minister in Athens tried for several months to induce the Greek Government to surrender Insull without success, but now it seems the Greeks have grown tired of Insull and according to a recent Reuters report from Athens, the Chicago promoter has been told he must leave Athens before January 31st. Apparently the Greek Government did not wish to permanently prejudice its friendly relationship with the United States through providing protection to a man who obviously had violated American laws through the sale of worthless stocks and bonds. Since Insull's passport has been cancelled he is in a serious predicament and according to officials of the State Department, he ultimately will be compelled to return to Chicago to face a grand jury indictment for using the mails to defraud the public in connection with vast money-juggling schemes.

"But while steps are being taken to return Insull to the United States for trial, nothing further has been heard for many weeks about C.C. Julian, another promoter of worthless stocks, who is residing in Shanghai in apparent safety against arrest by the American authorities. Like Insull, Julian was born a British subject, but Insull apparently became an American citizen while Julian retained his Canadian nationality. It is due to Julian's Canadian citizenship, that he apparently is free from molestation under exterritorial jurisdiction in Shanghai, for so long as he resides here neither the Americans or Chinese can arrest him and the British authorities claim they likewise are precluded from taking action because Julian's alleged crimes were not committed under the British flag, but in the United States. Just how much money Julian managed to store away before he fled from the United States is not known, but there apparently are plenty of people in California and Oklahoma who would like to get hold of him. A report from Los Angeles

< 118 >

which was published in the Kansas City Star on November 7th stated that losses from Julian's activities in that city amounted to well over $13,000,000 in U.S. Currency. The report was based on an investigation into federal receiverships which is now being conducted by a commission of five United States Senators. The report stated that Julian received $500,000 for his stock in one concern before it went into receivership. The claim was made that the receivers of the Julian Petroleum Company, though refusing to disclose the names of valid stockholders had prevented filing of suits for the recovery of losses up to $13,000,000. Serious charges were made by the Senatorial Commission that federal receivers received fees running from $210,000 to $390,000. It was also charged that there had been an over issue of 4,000,000 shares in one of Julian's fiascoes. Joseph Scott, who nominated Herbert Hoover for the Presidency in 1932, received $105,000 in receivership fees before he resigned the position in 1929, according to disclosures by the Senatorial investigation. All of which would seem to point to the disability of the American and British authorities getting together and working out some method whereby C.C. Julian could be returned to the United States in order that the whole story of his questionable practices could be brought out. At any rate no particular honor accrues to this city as a result of granting Julian refuge!"

Bill Feeney parked his Dodge in the driveway of his home. Gypsy greeted him barking and snapping at his car's front tire as it came to a halt on the incline of the driveway. Bill Jr. greeted his father at the front door.

"Hi Toluca Poluka, where's your mother?"

"She got a call from Universal to do a bit part this afternoon. Grandma is taking care of Jennifer."

"Hi Grandma, how's Jennifer?"

"Fine Bill, she eats a lot."

Bill picked up his daughter, sat her on his knee, and sang, "Toluca girls are the best in the West. They keep things going and they never take a rest. And they have one song and they sing it all together. It goes like this, 'Toluca girls forever'!" Jennifer giggled and clapped her hands. His wife soon entered the living room from the kitchen.

"Hi honey."

She walked over to Bill and kissed his cheek.

"It was an easy shoot over at Universal. But every penny counts."

Kathryn picked up her daughter, kissed her and put her down to toddle after her toward the kitchen. Bill Jr. let the dog in by the back door and went to his room. Bill Feeney poured a drink, casually pacing back and forth in the kitchen area while Kathryn and her mother prepared dinner. Kathryn's father had died the year before. Her mother had been staying with the Feeney's for the past month, intending to return East after New Years, 1934.

"Kathryn, I want to ask you and your mother something. Now don't you two get excited. Hank Bond called me to his office today. He thinks I'm doing a good job and he wants to reward me with a special assignment."

"What kind of special assignment Bill?"

< 119 >

"Show business."

"Isn't that de-ja-vu. I'm your wife of twelve years. You live near the studios. Surely he doesn't want you to do a gossip column does he?"

"No, he wants me to go to China."

"To China, good grief what for?"

"Metro Goldwyn Mayer is sending Director George Hill, some technicians and camera men to Shanghai to seek permission of the Nangking government to film "The Good Earth" in China. The MGM publicity department contacted the major metropolitan dailies in Los Angeles and New York offering to pay the fare for one reporter from each paper round trip to Shanghai to cover the negotiations. Hank Bond wants me to go, all expenses paid. I told him I'd have to discuss it with you before I give them a final answer."

"How long would you be gone?"

"Tops, two months. The group will leave for Shanghai on one of the Dollar Line ships between Christmas and New Years."

"Bill, we've never been apart for that length of time. I don't like it."

"What do you think, Grandma?"

"Bill, I'll stay and help Kathryn. The trip could mean a promotion for you."

"Kathryn, you awake?"

"What time is it?"

"Two, I can't sleep."

"What's the matter?"

"It's this China thing. We've never been separated since we met at Musso & Frank's and I don't think I can handle being away from you for two months. How about coming with me? We have enough money saved up. The hotel room in Shanghai for two can't be that much more than for one. Tomorrow ask your mother if she will take care of the kids. Do you think that's asking too much?"

"I don't think so, Bill. ll talk to her in the morning."

"Pull down your nickers."

"What for?"

"You know what for?"

"Then you pull them down. Do you have your condoms? I can see you're ready for one?"

"You bet and how about sex at sea. That'll be a new experience."

"If you're a good boy, we can have our third child conceived somewhere out in the Pacific.

"Yes, but let's wait for a sunset, a calm sea and the throbbing of the ship's engine to do it right."

"Bill Jr. was conceived during that weekend at Warner Hot Springs. Do you connect begetting children with the flow of water?"

"You know, a child swims for nine months before it emerges into this world of ours."

"No more talking Bill."

< 120 >

Bill and Kathryn Feeney mingled with other passengers on the deck of the Dollar Liner President McKinley. Their leave was delayed by Chinese military officers who supervised the unloading of a Curtiss-Condor bombing plane, a type never brought to China before. The plane was to be placed on a truck and taken to the Lunghwa Airfield. The twin engine plane bore three machine guns. According to Commander Frank Hawks, a retired U.S. Naval pilot who accompanied the plane, it was capable of dropping a ton of bombs. Finally the plane unloaded safely, the passengers of the President McKinley filed down the gangplank to the dock.

Bill and Kathryn took a taxi to the Park Hotel.

"That New York Times interview with Chauncey Julian took place in the dining room of this hotel. I wonder if C.C. is still here."

"Just keep him away from me, I don't want to see that braggadocio again."

"Kathryn, be polite if you bump into him."

They checked in with the reservation clerk.

"Sir, is there a C.C. Julian registered here?"

"No, Mr. Feeney, Mr. Julian checked out after Christmas. If you see him, will you remind him he forgot to pay his bill."

"Certainly, if I see him."

"Thank goodness he's gone. He's probably not far away. I bet he's broke."

"It's possible, Kathryn."

"Sir, will the hotel arrange for our luggage to be brought from the Customs House?"

"Yes, Mr. Feeney, it will be placed in your room before sunset."

From their window they had a view of the Soo Chow Creek. The Shanghai North Station was visible in the distance. Bill put his arm around Kathryn's waist. His hands moved upward clasping her breasts. He kissed her on her lips, "Well, here we are in Shanghai, Kathryn. No kids."

"I miss them, Bill."

"Would you like to take a walk along the Bund before dinner?"

"No Bill, I'm going to rest. You go out and get the lay of the land. Do you have to work tomorrow?"

"I don't think so. The rule is that they will leave a message at 7:00 a.m. with the hotel clerk when they want us. There is some kind of a controversy about Mrs. Buck's novel. Some Chinese object to her allusions to opium smoking, concubine taking, and brigandage as being prejudicial to the prestige and dignity of the Chinese people. That problem has to be ironed out before they talk about shooting the film. Why don't you tidy up. I'll be back in an hour or so."

Bill Feeney walked over to the Bund and dropped by the North China Daily News and presented his press credentials. He was told to stop by the police station on Nangking near the hotel the following day to get a press pass. Bill then entered the long bar of the Shanghai Club and walked the length of it. Several of the reporters and staff brought over by MGM were at the bar as well as Japanese and American naval officers.

< 121 >

"Bill Feeney, welcome to Shanghai! What brings you to this field of infinite opportunity?"

"Hello C.C.."

The two men shook hands.

"Have a drink, Bill. On me."

"Thanks C.C.. I've been reading about you. I guess you're safe here."

"Safe and sorry, Bill. Safe and sorry. They're not there any more, Bill."

"Who C.C.?"

"My folks, the little guys. There are no folks to talk to. The big shots here don't trust me."

"Don't you have any friends here, C.C.?"

"Not many...one wonderful woman, my secretary Leonora Levy, one or two others. Is Kathryn with you?"

"Yes, she's resting at the hotel."

"Does she ever hear from Mary Olive?"

"Yes, they correspond. It seems the two girls are doing well in school. Your oldest daughter is in nursing school. By the way, C.C., if it's any consolation to you, Standard Oil, Signal Oil of California, and Signal Oil of Delaware are under investigation in Los Angeles for violation of the National Petroleum Administration Code of Competition by selling under posted prices. The word is out that the grand jury is going to indict the companies and many corporate officers."

"Believe me, Bill, nothing will come of it. Those old boys in Los Angeles, they take care of themselves. In the Julian Pete mess, that D.A. Asa Keyes took it on the chin worse than the rest of those pool boys and he, eventually, was pardoned by the Governor. How long are you going to be in town?"

"Well be leaving by the end of March. We're scheduled to leave on the twenty third. Any messages for folks back in California?"

"Yes Bill, you can tell Amy Semple McPherson and Death Valley Scotty that the formula didn't work for me in China."

"What formula?"

"They'll know. See ya later, Bill."

"So long, C.C."

Julian tossed a two dollar bill on the bar, smiled and eased away stiffly.

Chauncey Julian walked briskly from the Shanghai Club en route to his hotel in the French quarter. As he passed the American Club, J.F. Malone, the American business representative whom he met at the bar at the Shanghai Club his first night in Shanghai called out to him.

"C.C., I've looking all over for you. There's a man here from Oklahoma who wants to talk to you. Come over to my office and I'll phone him at his hotel."

"Is he a federal agent?"

"No, his name is Roy Hoffman. He said you would remember him."

< 122 >

Julian followed Malone to his nearby office where Malone placed a phone call to the Park Hotel. In ten minutes Major General Roy Hoffman of Oklahoma City entered Malone's office. He shook hands with Julian.

"What can I do for you Roy?"

"Look C.C., this is no place for you. The President Jackson is sailing for the states on the twenty third. Come back to Oklahoma and face trial. They may have lost their witnesses by now. I'm willing to pay for your fare."

"Thanks, Roy, but since coming to China I've basked in the sunbeams of idleness. Now I am one hundred percent alive. I can't fold my hands for the remainder of my life. Therefore Shanghai is home and I'm going to work here. Besides, I couldn't get a fair trial in Oklahoma. Alfalfa Bill is the tool of the large oil operators. Also, old Alfalfa is politically dead and he's trying to get his name in the newspapers by hounding me."

"Well C.C., I think you're making a mistake. If you change your mind call me at the hotel or call Mr. Malone here." The two men shook hands.

"Goodbye, Roy."

"Goodbye, C.C."

"Thanks for the use of your office, Malone."

"Anytime."

Julian returned to his hotel room where Leonora Levy was waiting for him.

"Let's have dinner, C.C."

"Surely Leonora, let me freshen up."

Leonora put on her coat. It was chilly outside and the couple walked arm in arm towards a small French restaurant down the street from the hotel.

"You know Leonora, I read an article recently by a Chinese journalist. He complained that blackmailers and suicides are becoming common place and that newspapers, by playing up the incidents, are encouraging others to imitate this behavior. The article implies that the papers tend to glorify those who perish at their own hands. Did you read the article?"

"No, C.C."

"The magazine is back at the hotel. Remind me to give it to you."

"I don't want to read about blackmail or suicide, C.C."

"Shall we dine?"

The following day, unannounced, Chauncey Julian appeared at the office of J.F. Malone.

"May I have a word with you, J.F.?"

"Sure C.C., come on in. Take the load off your feet. What can I do for you?"

"J.F., I've completed my memoirs, 'What Price Fugitive?' I would like to raise some cash by selling a partial interest. Could you locate someone with some capital to invest who would be interested?"

"Its a possibility, C.C.. Let me take a look at the manuscript."

"That's not possible. I mailed it to a New York publisher."

"Did you keep a copy of the manuscript here in Shanghai?"

"No, its all in my head."

"All right, C.C., I'll look around. I'll call you."

< 123 >

"Thanks for your time, Malone."

Julian returned on foot to his hotel room in the French Quarter and Leonora arrived at the same time in a rickshaw. Julian helped her alight and kissed her on the
cheek.

"How's Ralph?"

"Grumpy, C.C., he complains he sees so little of me anymore."

The two entered the hotel. Julian picked up his mail.

"Did you hear from the publisher yet, C.C.?"

"Not yet, my dear."

The couple walked to Julian's room.

"Leonora, Saturday will mark one year for us here in Shanghai. I'm going to have a party, just for the two of us at the Astor House. I've taken a room and arranged for a sumptuous dinner. How does that strike you?"

"On, I think that's wonderful, C.C., I know just what to wear."

Leonora borrowed her brother's Ford for the evening. She met Julian at his hotel and the two drove to the Astor House, a fashionable downtown tourist rendezvous. Julian and Leonora were escorted to their table. Julian greeted several friends and acquaintances. J.F. Malone was there. Dr. C.Y. Whang and his wife nodded recognition. George R. Coleman of New York, Vice President of Elbrook Incorporated, shook hands with Julian. A Russian woman, the companion of an obviously wealthy Chinese businessman caught Julian's eye. He remembered their encounter his first night in Shanghai. He saluted her with a wave of his hand. She smiled faintly.

At their table Julian and Leonora were served an elegant dinner of both Chinese and American cuisine. Shortly after midnight, Julian turned to her.

"Leonora, excuse me, I need to freshen up a bit and will be back in a few minutes."

He gently grasped Leonora Levy's left hand. He bowed, kissed her hand, then her lips and moved swiftly from the crowded dining room area toward the living quarters at the Astor House.

Leonora thought back over the past months. Her love and admiration for Chauncey Julian had grown, not subsided, despite Julian's failure to establish himself in the Shanghai business community. Her loyalty to Julian was an offshoot, not only of her genuine love for the man, but also from her conviction that C. C. Julian in the past had been a great American capitalist victimized by sources of political corruption linked with greedy bankers, both in the states and more recently in Shanghai.

"Miss Levy, J. F. Malone. Mr. Julian spoke quite highly of you. You are his secretary, are you not?"

"Yes, Mr. Malone."

"May I have a word with you?"

"Of course, Mr. Malone. Mr. Julian has excused himself for a few moments."

"Miss Levy, Mr. Julian approached me the other day concerning his memoirs. I sent a wire to a friend in New York. He represents authors

< 124 >

negotiating with publishers. He expressed an interest in representing Mr. Julian. Please ask Mr. Julian to give me a ring."

"Certainly, Mr. Malone. He knows your number?"

"Here is my business card. Have Mr. Julian call me next week. Good evening, Miss Levy."

J. F. Malone returned to his table. Leonora Levy checked her wrist watch—12:15 a.m. She turned toward the maitre d'hotel with a quizzical expression. The maitre d'hotel approached her table. "Madam?"

"I suppose it's nothing. He usually..."

"Miss Levy." A buxom Russian woman glided her ample body into Julian's chair. She guided a gold and ivory cigarette holder to her lips exhaling an intended smoke ring toward the startled Leonora Levy. "Darling, I am Mary Cantorovich, an old friend of Mr. Julian. My dear, let me get to the point. I have a dear friend who is a manager of a well known tavern here in Shanghai. He has been approached by an important business man from Oklahoma. Let me be blunt, Miss Levy, people, important people know that C. C. Julian is, how do you say it in English, bust, no, broke. There is a rumor that Julian wrote a book. His memories, I believe he called it. There are people willing to pay Julian a tidy sum of money not to publish his book. Tell Mr. Julian I want to see him. He knows my friend at the tavern. Tell him to come to me tomorrow night at 10:00 p.m. Now excuse me, Miss Levy. Good evening."

Leonora again looked at her wrist watch. She became apprehensive. Julian had never before left her alone for so long.

"Waiter, I'm concerned about Mr. Julian. He never leaves me like this. I'm going to his room to see if he is feeling ill."

Leonora walked hurriedly to Julian's room and knocked at the door. There was no answer. The door was unlocked. Opening it, Leonora saw Julian slumped in his chair. He appeared to be unconscious. A small bottle, almost empty, was on the desk with a glass stopper nearby. Leonora rushed to the hotel front desk.

"Come quickly, something has happened to Mr. Julian. He's unconscious."

Leonora and the clerk ran to Julian's room. The clerk shook Julian. Julian tumbled. The hotel attendant called to his assistant.

"Quickly, call the house doctor."

The physician arrived, checked Julian's pulse and heartbeat.

"Summon an ambulance."

Attendants rushed Julian to the hospital and Leonora followed in her brother's Ford. At the hospital, Leonora paced back and forth near Julian's room.

"I'm sorry Miss, Mr. Julian has passed away."

Sobbing, she ran to the Ford. With a blank stare she drove back to the Astor House. Forlornly, she walked to Julian's room, grasped the bottle of poison, swallowed all that remained and collapsed on the floor.

"Leonora, this is Dr. Yujen. Can you speak to me?"

"Julian said he would do it. I didn't believe him but

< 125 >

he did it. He was a brave man."

"Leonora, where did Julian get the strychnine? She's unconscious."

"I demand to see the body of Chauncey Julian! He is my dear friend." A woman with a Russian accent screamed at the nurse on duty. Hospital staff surrounded her as she gestured weeping and muttering words in Russian. I am Mary Cantorovich. I demand to see Julian."

"Miss Cantorovich, Mr. Julian is dead."

"I demand to see the body."

"I am sorry but that is forbidden by the coroner."

Hank Bond walked over to Bill Feeney's desk at the Citizen News.

"This just came in over the Associated Press Wire Service. Chauncey Julian is dead. Suicide."

"Oh, no! He bought me a drink less than a month ago. I wonder if his family knows."

"Does he have any family here in town."

"Yes Hank, a sister in Hollywood."

"Get over and interview her and hire a stringer in Shanghai to cover the funeral."

"I met a fellow when I was over there. I'll send him a wire. I'm sure he will follow through for us."

Feeney returned to his desk and picked up the phone.

"Maggie, try to get my wife on the phone. She should be home. I'll be dropping a telegram on my way out. Phone it in for me please."

"Hello, Kathryn, we just received a wire service report that C.C. Julian took poison and died. Would you call Mary Olive? It's possible no one contacted her."

"I will, Bill. I believe she knew he would end up like that."

"That's not for me to say, Kathryn, I've got to go to his sister's home now. I'll talk to you later."

Feeney drove his Dodge to the home of Violet Greenshaw on Franklin Place in Hollywood.

"Good morning, Mrs. Greenshaw. My name is Feeney from the Citizen News. I'm inquiring if you have been informed about the situation in Shanghai."

"I have not heard anything from Shanghai, Mr. Feeney."

"May I come in?"

"Certainly."

"Well, I regret to inform you, Mrs. Greenshaw, that your brother C.C. Julian passed away. He took his own life."

Violet Greenshaw seated herself on a sofa.

"Has the widow been notified, Mr. Feeney?" "Yes, I took care of that. Obviously you're not aware of the place for the funeral."

"No, I don't know whether the widow will make any arrangements."

"Do you know who the executor of his estate is, Mrs. Greenshaw?"

< 126 >

"Yes, I am the administrator, Mr. Feeney. The assets of the estate consist of some promissory notes with a face value of five thousand five hundred dollars, accounts receivable, some personal property, oil royalties and some producing wells. I'll file a letter of administration of the estate sometime next week."

"Thank you, Mrs. Greenshaw, and my condolences to you and C.C.'s brothers. I knew C.C. better than most people in this town. There was a lot to admire. May I use your phone, Mrs. Greenshaw?"

"Certainly, Mr. Feeney."

"Information? May I have the phone number of the United States Postal Inspector's Office in Burbank, California?"

"Operator. Operator Sunset 6900."

"May I speak to Postal Inspector Madeira? Bill Feeney from the Hollywood Citizen News."

"Just one moment."

"Hello, Feeney. Do you have a problem with the mail?"

"No, Madeira. Your career problem with the mail is no more."

"What are you driving at?"

"C.C. Julian is dead. Suicide in Shanghai."

"That's one way to close the file. You know, Feeney, if he had returned, all he would have faced was a bail jumping charge. I lost both my witnesses—one dead in an automobile accident and the other fellow became a buyer for the Broadway Department store in England and as much as told me he could refuse to answer a summons."

"Well, Madeira, you can retire knowing that Lady Justice, although blindfolded, was served."

"Mr. Feeney, this just came in by wire service from Shanghai."

C.C. IS BURIED

(Special Bulletin - Hollywood Citizen News)
Shanghai, May 11th.

C.C. Julian was buried here today. Graveside services were performed by the Reverend Emory Luccock, Pastor of the local American Community Church. The burial location was a small cemetery for foreigners just outside of Shanghai. The sole mourner was Miss Lenora Levy dressed in white.

< END >